DATE DUE

REINVENTING MAN

REINVENTING MAN

The Robot Becomes Reality

Igor Aleksander and Piers Burnett

HOLT, RINEHART AND WINSTON New York

First published in the United States in 1984 by Holt, Rinehart and Winston, 383 Madison Avenue, New York, New York 10017.
Published simultaneously in Canada by Holt, Rinehart and Winston of Canada, Limited.

Library of Congress Cataloging in Publication Data
Aleksander, Igor.
Reinventing man.
Bibliography: p.
Includes index.
1. Robotics. 2. Artificial intelligence.
I. Burnett, Piers. II. Title.
TJ211.A43 1984 629.8'92 83-18388
ISBN: 0-03-063857-7

First American Edition

Printed in the United States of America
10 9 8 7 6 5 4 3 2 1

ISBN 0-03-063857-7

Contents

Preface

Robots are very much in vogue. Indeed, the developed nations have embarked on something of a 'robot race' in which industrial virility is measured by the number of robots installed in their factories, much as Dreadnought battleships were used as a yardstick of military might and diplomatic resolve in the decade preceding the First World War.

We hope that a robot race is a symptom of a world which has grown saner than the one which embarked on the battleship race, but it would be absurd to suggest that robots represent a totally benevolent technology. The growth of the robot population, which is only one facet of the spread of automation in general, is already costing human jobs and will, no doubt, cost more in the future. The fact that this book refers only in passing to the social and economic issues raised by the development of robotics should not be taken to mean that we find such problems uninteresting or judge them unimportant. On the contrary, we consider them so significant that we have preferred to omit discussion of them rather than allotting them a subsidiary or secondary place in a book which is designed to concentrate on other aspects. It may, however, be worth adding that no one should conclude that, because we favour models of human behaviour and intelligence which do not rely on preprogramming, we necessarily believe that all problems are best tackled in an unprogrammed, *laissez-faire* fashion.

Even though we have chosen to discuss the robots of today and those of the future mainly in terms of how they do things rather than what they do, we do not think that this is irrelevant to those who have to consider the jobs that will be lost and the skills that will be supplanted. For, when dealing with a technology which, although barely two decades old, is already having a major impact on industry and employment, it would seem essential for those whose concern is with its social and political consequences to understand the technological factors involved. How can an economist or a sociologist assess the future effect of robotics in a particular field unless he or she has some grasp of how robots are developing, current trends in research, and the advances that are most likely to be made? We hope, therefore, that this *tour d'horizon* of contemporary robotics will help those whose business it is to plan for a future in which industrial robots will play an ever-increasing role.

But, as will become increasingly clear to the reader as the book progresses, our principal interest is not the robot as a utilitarian machine or a source of profit. Rather, we are mainly concerned with the idea of the robot as an image of man. We are well aware that, in presuming to compare men and machines, we are stepping into turbulent philosophical and metaphysical waters. We make no apology for this but leave the reader to judge whether the endeavour has proved worthwhile.

One final point. In order to make our case – that men and machines can be compared both validly and profitably at some level, though not at all levels – we have had to explain in some detail the workings of quite a lot of machines. We hope that we have succeeded in doing so in a way which will neither offend the *cognoscenti* by appearing simplistic or inaccurate, nor, even more importantly, rebuff the general reader by giving the impression that the subject is arcanely technical and, therefore, 'difficult'. Though we have avoided jargon and anything more demanding than elementary mathematics, we have above all tried not to dodge the issue by glossing over important technicalities with vague and ill-defined generalities. Robots are, it seems to us, not so much too important as quite simply too interesting to be left to roboticists.

IA

PB

July 1983

1. The Mechanical Doppelgänger

Though our smoke may hide the Heavens from your eyes,
It will vanish and the stars will shine again,
Because, for all our power and weight and size,
We are nothing more than children of your brain.

Rudyard Kipling, 'The Secret of the Machines'

In the early days of the Second World War the British government set up a committee to vet the various suggestions for war-winning weapons which flooded in from inventive citizens. The story is told of one such proposal, for a giant airship armed with death rays. Its inventor was summoned before the committee and produced blueprints of his vessel, its engines and navigational systems. He was just launching into a description of the protective suits he had designed for the crew when the chairman of the committee broke in, impatient to get to the heart of the matter, 'But what about the death rays?', he inquired.

The inventor was taken aback. 'Oh,' he said, 'I thought you had lots of *those*.'

If a robot is defined, provisionally, as 'a machine made in the image of man', then it has to be said that, like death rays in 1940, robots in 1983 remain in the province of speculation rather than established fact. But, even if we do not share the inventor's tendency to confuse the two categories, we do seem to have accepted the advent of the robot as inevitable, a question of 'when?' rather than of 'if?'. Even more significantly, just as the inventor took the existence of death rays for granted and concentrated on devising a method of bringing them into action, so the popular imagination has been much more preoccupied by the question of what we should do when robots *do* arrive on the scene than by the business of actually making them. It is rather as if the Wright brothers' first flight had been preceded by an extensive literature on air traffic control.

The prevailing image of the robot from which we foresee so much trouble is probably that of a walking, talking 'mechanical man', and this concept has become so established in our consciousness that it takes an effort of will to remind ourselves that we are no more likely to encounter anything that matches up to it in reality than a medieval knight would meet a hippogriff face to face. We are dealing with hypothesis rather than fact. The first interesting question is: why and how did the hypothesis take on so vivid a form?

There are, it is true, machines in existence which we have, perhaps rather prematurely, categorized as robots. But long before even this primitive vanguard became a practical proposition the idea of the robot was enjoying a lively existence in the human imagination. It seems to have been around even before it gained a name. Other inventions had to take concrete form before a name was found for them, and it took some time for the English speaking world to agree that the 'horseless carriage' should be a motor car or a 'flying machine' an airplane, yet when Karel Capek published his play *RUR* (for 'Rossum's Universal Robots') in 1923, the obscure word (it means no more than 'serf' or 'worker' in Capek's native Czech) was quickly and universally adopted. Capek's 'robots' were but the latest in a whole line of mechanical men, and all that the concept had hitherto lacked was a label.

Sixty years later, while their real-life counterparts are only just undergoing their birth pangs, fictional robots have become as familiar a part of our imaginative lives as cowboys and Indians or cops and robbers; many a hero of contemporary space opera would be as lost without a robot companion as the Lone Ranger without Tonto or Holmes bereft of Watson. Moreover, from a very early stage the new genre of science or speculative fiction had seen that the robot had possibilities that most of its other technological paraphernalia lacked. Spaceships and time travel served to transfer old plots to new planets, but robots, along with those other novelties, alien beings from distant worlds, were jokers that could be dealt into the existing pack to produce a whole new game. The robot was not a prop, it was a character, and one which could be used to explore new metaphysical issues. Those exponents of sci-fi who aimed at simple entertainment were content to use the robot to play Watson to a human Holmes, but more serious writers saw that it had far greater potential in the role of Holmes himself – or, perhaps, Moriarty.

We began with a working definition of a robot as 'a machine made in the image of man'. Although it is already evident that, in purely physical terms, this description is likely to prove misleading, much science fiction continues to utilize it if only because it effectively dramatizes a profound metaphysical issue. The robot's status, somewhere in an uneasy limbo between man and machine, makes it the ideal dramatic device for exploring the relationship between the two. The robot can conveniently stand for the whole of machinekind which twentieth century man has increasingly come to see as a potential threat, or even successor, to mankind. Writers from Capek onwards have used the robot in this way to explore the possible forms which man's final confrontation with the machine might take and to forecast likely outcomes – few of which have been encouraging for the human race. At best, in basically optimistic writers like Asimov, the robot is seen as fundamentally superior to human beings ('a better, cleaner race') but prepared to exercise a benevolent dictatorship over them; at worst, Capek and his even more pessimistic successors have envisaged robots as the evolutionary successors to their human creators who are fated to go down to extinction like the dinosaurs.

Even the less portentous kind of science fiction has capitalized on the robot's potential as a 'baddie'. In *Star Wars*, for example, the amiable, twittering 'droids' may seem to be making a bid for the role of 'spaceman's best friend', but the villainy of Darth Vader is certainly enhanced by a robotical aspect, and the same could be said of the Daleks in *Dr Who*; as early as 1898 in *The War of the Worlds* H G Wells foresaw an invasion of aliens who manifested themselves in the form of giant machines. Scientific developments in genetic engineering and bionics together, perhaps, with the problems of shoe-horning actors into plausible hardware, have spawned whole subspecies like androids, humanoids and cyborgs in which the man-machine ambiguity is emphasized by making the two physically indistinguishable so that the robots become not just an external enemy but a fifth column within the human ranks.

By pushing the cart of speculation so far ahead of the horse of invention and, indeed, suggesting implicitly that it might be best to slam the stable door on the horse rather than letting it out at all, science fiction has cast a pretty gloomy horoscope for the real robots which are the subject of this book. But before looking at why it might be wrong to burden the infant at birth with all the sins of its mechanical forefathers, it is necessary to go back and look at the taint of original sin it inherits from an even older predecessor, the 'unnatural man'.

Attempts to provide the robot with an ancestry usually go back to antiquity. The Greek myths contain a number of beings which could be viewed as in some way 'robotical'. Mechanical soldiers crop up in *The Odyssey*; Thebes was founded by a race which sprang from the dragons' teeth which Cadmus scattered on the instructions of the goddess Athene; another deity, Hephaestus, provided King Minos of Crete with Talos, a giant statue which guarded his kingdom; and a descendant of Hephaestus, Daedalus, was credited with a bronze warrior which confronted the Argonauts. It is not, however, clear that these examples can be gathered into a distinct category, except with hindsight, or that they were seen at the time as anything out of the ordinary in a universe which contained Medusa, centaurs and minotaurs. Indeed, such beings seem an almost inevitable product of a pantheon which intervened arbitrarily in human affairs on a day-to-day basis.

Perhaps the most widely quoted example of a 'classical' robot is Galatea, the consort of Pygmalion. Pygmalion was a young man who was ambitious enough to fall in love with the goddess Athene, a passion which was unrequited, in physical terms at least. To console himself he fashioned a statue of his beloved which he took to bed with him. Touched by this evidence of his devotion, but still unable to oblige in person, Athene provided a surrogate by bringing the statue to life as Galatea. In our view, it is stretching a point to call Galatea a robot, for it is of the essence of the robot that it is man-made. The theme has, however, been picked up in contempary science fiction by several writers who have provided their Pygmalions with mechanical Galateas. The same sort of reservations apply to the 'mechanical men' which crop up in other literatures; they are products of direct divine intervention, or of a harnessing of supernatural forces through sorcery; and, though they are mysterious and frequently inimical to the human characters, they hold no more terror for a competent hero than any of the other perils he encounters.

It seems to have been from the Judeo-Christian tradition that 'unnatural man' received his unique charge of superstition. It is not surprising that a single God, and one who had made man in His own image, should take the creation of life a good deal more seriously than the frivolous gods of the Greeks. Jehovah forbade his followers graven images and it is inconceivable that he would have brought one of them to life simply to gratify a devoted follower. On the other hand, if God was the source of all life, and man could invoke His powers through prayer and ritual, it was not impossible that the means of bestowing life were within the grasp of the pious and learned. The most influential product of this line of reasoning was the *golem* of Jewish legend, a clay figure endowed with human features through the performance of ritual and, ultimately, brought to life when the name of God was inscribed on its forehead.

Many accounts credit rabbis with creating *golems* to serve in their households, a sort of magical race of gentlemen's gentlemen whose only drawback was that the power of speech was withheld from them. But a much more dramatic version of the legend is associated with Rabbi Loew of Prague who is alleged to have created a *golem* in order to protect the Jewish community in the city against one of the recurrent pogroms. In fact, later research has established that Loew was a sober theologian who would no more have meddled with *golems* than serve roast pork at a *bar-mitzvah* and the *golem* has moved further back into the realms of conjecture. But the *golem* attributed to Loew had two features that ensured its survival as a theme in fiction – it was superhumanly powerful (how else could it deter the gentiles?) and it was prone to get out of control and turn upon its creator. In one version of the story the creature ultimately runs amok and its career of destruction is only ended when the rabbi tricks it into kneeling before him and plucks the magic formula from its brow, thus reducing it to clay again.

The *golem*, however uncontrollable it may have proved in practice, was always created from the best of motives and with the aid and sanction of God. Christianity, with its much greater emphasis on evil as an active force competing for men's souls in the world, has taken a much more jaundiced view of interfering with the life force than Judaism has. The two traditions coincide to the extent that in both the creation of unnatural beings involves the invocation of supernatural forces, but the tendency in Christian mythology is to suggest that those forces are diabolical rather than divine. There is no counterpart of the *golem* in Christian mythology, but the medieval world was much intrigued by the *homunculus* – essentially the minuscule but perfect human being which, medieval medicine held, was contained in the male sperm and implanted in a woman's body to grow into the foetus – has subsequently been quoted as a prescientific instance of robotic preoccupations. The effort to create homunculi artificially was a practice closely linked with alchemy, the transmutation of base elements into gold, another process which required supernatural aid. Both practices are associated with historical characters who have, at best, shady reputations; Paracelsus and Faust, for example. On balance, the image of such men which prevailed was that of scholars who rashly ventured into forbidden areas of knowledge, probably at the price of their own damnation. They were sorcerers, not saints.

Like many another concept, the idea of the unnatural man attained its definitive form after the beliefs that gave rise to it were discredited, and gained impetus by co-opting a new trend. Mary Shelley's *Frankenstein* might be called the paradigm version of the *golem* legend and much of its success was due to the skill with which its author leavened the mystical and the magical with more than a pinch of the new science. In order to assemble his creation, Frankenstein ransacked the charnel houses (a sinister enough procedure in itself) and delved back into forgotten lore to rediscover the secrets of the ancients; but the description of his procedures also draws upon early nineteenth century chemistry and though the author is, not unreasonably, a little reticent about 'the cause of generation and life', it seems to have had something to do with electricity – a phenomenon which was still far from understood in 1819. Frankenstein's monster was, of course, novel in one other respect: it was the creation of man alone, unaided by any supernatural powers, and this made the fact that it turned upon its creator all the more significant. Frankenstein was not meddling with powers beyond his comprehension; he was utilizing knowledge which he had taken immense pains to acquire and understand, but the consequences were no less disastrous and enormously more ironical.

Mary Shelley was, of course, a leading member of the romantic movement which rejected much of the rationality of the eighteenth century. One of the achievements of that rationality had been to substitute an orderly, explicable universe for a magical and supernatural one. A start had been made upon the task of classifying nature and the laws which governed its behaviour were being revealed; and it was even becoming conceivable that the workings of man himself could be comprehended in terms of rational cause and effect. In 1748 the French writer La Mettrie, in his book *L'Homme Machine*, had stated the philosophical case for considering man as no more than an elaborate machine. Not many would have gone as far as La Mettrie; Christianity required that the soul be reserved as the territory of the spiritual rather than the material and most opinion continued to make a distinction between body and mind, being prepared to cede the former to science while maintaining that the latter was outside its scope.

But to the romantics the new science, with its cold, reasoned methods, was as sterile as the classical Augustan literature against which they were avowedly in revolt. And *Frankenstein* turned the tables on the age of reason brilliantly. Even the grisly methods by which the monster was assembled reflected the contemporary reputation of anatomists as grave robbers and pillagers of corpses. The novel lumped the new knowledge together with the old, tarred them both with the same brush of sacrilege and turned the case for the new science on its head by suggesting that, rather than being a reassuring substitute for superstition, it would produce even more horrific consequences.

Much of *Frankenstein*'s power comes from its combination of the old idea that there were limits beyond which man explored knowledge at his own peril with the modern, romantic theme that human nature compelled him across that forbidden frontier. The plot also contains a novel psychological twist: the monster is Frankenstein's creation not only because he brings it into physical existence but also in the sense that it is his rejection of it that guarantees its malignancy. When it first comes to life the monster is gentle and well-inclined and it is only when Frankenstein realizes the enormity of his actions, and cannot conceal his revulsion at the outcome, that its nature is transformed. There is an interesting parallel here with the way in which the *golem* legend fared. Many of the original *golems* were dumb, harmless creatures content to perform useful tasks about the house, but popular taste ignored these in favour of what might be called the King Kong version. Obviously, there is something of Dr Frankenstein in us all, in that we expect unnatural man to reflect and magnify our vices rather than our virtues.

The robot's problems, it might be said, began with *Frankenstein*. The novel became the classical parable of scientific hubris, and the man-made monster the classic instrument of denouement. The physicists had to produce a bomb before, in Oppenheimer's famous phrase, they had 'known guilt'. The roboticists, the heirs of Frankenstein, were under suspicion before they even started work.

The history of the robot has also been traced back through a family tree much less sinister than that of the unnatural man. Unlike that of *golems* or *homunculi*, the history of automata is one of fact rather than myth. An automaton in the classical sense of the word (it has specialized meanings in contemporary mathematics and computer science which we will encounter later) is a mechanical device which mimics, in one or more ways, the outward appearance of life. Thus a clockwork mouse or even the *Manneken-Pis* statue in Brussels, endlessly relieving itself into an ornamental pool, are automata. Again, the tradition can be traced back to the ancient world, where the work of skilled exponents of the art like Hero of Alexander was celebrated in contemporary accounts. Their accomplishments, and those of their medieval successors, were exaggerated in the telling, no doubt, until they were credited with feats that were supernatural rather than merely ingenious. Certainly, the examples and designs which have survived do not suggest workings which would have mystified any reasonably sophisticated observer.

The eighteenth and nineteenth centuries brought the art of the automaton to a new peak. The refinement of the clockwork mechanism and advances in materials and precision made possible such marvels of verisimilitude as Jacques de Vaucanson's duck, which not only flapped its wings and quacked but also ingested and excreted food, or the clockwork dolls of the two Jacquet-Droz (father and son), one of which could write a short letter while another drew portraits of well-known people. By the latter half of the nineteenth century clockwork dolls, even entire theatrical performances by a cast of automata, had become commonplace and popular. Perhaps the most bizarre product of this craze was the 'steam man', the invention of a Canadian engineer, which strode round in circles fuelled by a boiler in its abdomen and puffed smoke from a simulated cigar.

It is not, however, clear that the makers of automata saw themselves involved in an attempt to realize in practical form the ideas of materialist thinkers like La Mettrie. Much of the appeal of automata lies in the element of trickery involved; the audience is perfectly aware that what they are seeing is an ingenious illusion, and the effect is entertaining rather than disconcerting. This quality of innocence is reflected in the fact that automata inspired no equivalent of *Frankenstein*. There are, it is true, several stories like *Pinocchio*, *Coppelia* and Hans Christian Andersen's 'The

Emperor's Singing Bird' which hinge upon the idea of toys or automata coming to life. Some of them, *Pinocchio* for example, explore the idea of the inevitable alienation of unnatural man from his natural counterpart, very much a part of Mary Shelley's theme. But, significantly, they all start from the fairy tale's premise of 'let's suppose', which invites the reader to suspend disbelief. *Frankenstein*, in contrast, challenges the reader to dismiss the story as fantasy if he dares, by taking a typically modern standpoint and asking 'What if . . . ?'

Automata have certainly not contributed anything to the actual science of robotics, which could not have come into existence without techniques and ideas unknown in the heyday of the automaton. And it does not seem that they played a large part in bringing the idea of the robot into being, for the automaton has always been based upon the idea of the machine as toy or entertaining curiosity whereas the robot had built into it from the beginning the fear of the machine as enemy or competitor to man.

The robot proper seems to have been 'invented' some time around the beginning of the twentieth century for a combination of reasons. The success of writers like Jules Verne and H G Wells had popularized the new kind of speculative fiction which was set in possible futures rather than the familiar present or the mythical past, and the enormous proliferation of new inventions in the final decades of the nineteenth century seemed to justify novelists in taking bold leaps rather than cautious steps when examining what those futures might hold. Darwin's *Origin of Species*, and more particularly *The Descent of Man*, had reinforced the idea that the phenomenon of man must be explicable in materialist terms, and it was tempting to suppose that what nature had achieved by accident science should ultimately be able to reproduce by design. But even in the heyday of the idea of progress, the reaction to scientific advance was not universally optimistic. There was a widespread feeling that industrial civilization, and the technologies that made it possible, posed as many problems as they solved and, since disaster has always been a more remunerative fictional theme than Utopia, there was a considerable incentive to explore the possibility of man's love affair with the machine ending in tears. Finally, with 'unnatural man' lying to hand, already charged up with a very powerful set of associations, it was almost inevitable that the *golem* and

Frankenstein's monster would be pulled out of the lumber room of literature and refurbished in robot guise.

In truth, though the new genre of speculative fiction licensed writers to extrapolate existing scientific trends far into the future, there was little in late nineteenth century science, and certainly not any potential for development in existing automata, which pointed towards the possibility of real robots. The devising of a credible model of the brain and of a means of linking it with the body and the senses, which have emerged as the central problems of robotics, still lay well in the future. What was evident was a clear progress towards the mechanization of many departments of life – industry, commerce, communications, even the household – and a growing suspicion that this was leading towards a domination of machine over man.

Some embraced the trend. In the early decades of the century the Futurists proclaimed that the machine would create a future for humanity that would be quite different from, and much better than, its past. Thus Futurist writings glorified the machine and their paintings celebrated the partnership between man and machine, or even the symbiosis of the two in beings that were, at least symbolically, robots. Even if they had not the faintest idea of how it was to be achieved, advocates of the machine saw the robot as its logical epitome.

So, too, did those who saw the machine as dangerous and interpreted its growing influence as a threat. It was patently absurd to credit existing machinery with any conscious intent, but the robot conveniently represented both a hypothetical machine of the future and one which could reasonably be credited with a capacity for deliberate aggression. The fact that it had been made by man in his own image, and thus had resonances with earlier unnatural man, emphasized the irony. Capek was but one of many to see the dramatic possibilities of this scenario. In *RUR* the robots eliminate their human creators, until, when one final survivor remains, they belatedly realize that they have failed to master the secrets of their own construction. Unfortunately, the remaining human is no roboticist and can only acquire the necessary knowledge by dismantling, or dissecting, a representative robot of each sex. At the last moment, however, the two victims develop a romantic interest in each other and the audience is left with the vision of them as a new Adam and Eve.

Naturally, neither Capek, nor anyone else, for the next twenty years at least, had an inkling of how a real robot might be made or what it might look like. Rossum's robots did not aim at anything other than a symbolically mechanical quality and are played by human actors – modern science fiction would classify them as androids. In the movies it was different. Robots were an irresistible challenge to the special effects department and during the 1920s and 1930s a whole succession of mechanical men clanked across the screen. The hardware mostly appears to have been based on diving suits with radio aerials distributed *ad lib* around the helmet, though Fritz Lang's film *Metropolis* – which portrayed the whole mechanical kingdom in an apocalyptic light – featured a female art deco robot which prefigured C3PO in *Star Wars* to a startling degree – they would make a touching couple in the final scene of *RUR*.

The first writer to take robots seriously enough to try and contrive a plausible scientific background for them was Isaac Asimov. The robots in his prescient, and now classic, collection of stories, *I, Robot*, the first of which appeared in 1941, have 'positronic' brains. The workings of these are a mystery, even to their builders (not surprisingly since the computer was scarcely a twinkle in a mathematician's eye at the time) but they are sufficiently complex to require the services of a 'robopsychologist' to sort things out when they go wrong. Asimov foresaw many of the themes that were to preoccupy later writers: his very first story deals with the problems of a human developing an affection for a robot and the final one with a robot which is indistinguishable from a human; others concern robots which develop megalomania or become paranoid liars. He also foresaw at least two problems that have already cropped up in reality – his robots are excluded from the earth because of the competition that they would offer to human workers, and one of them is incapacitated by what a modern roboticist would recognize as a 'limit cycle'.

But the contribution of *I, Robot* which is best remembered and most frequently 'borrowed' by other writers are 'Three Laws of Robotics'. These are programmed into the positronic brains of all robots as a sort of unbreakable moral code. The three laws are as follows:

1 A robot must not harm a human being, nor through inaction allow one to come to harm.

2 A robot must always obey human beings, unless that is in conflict with the first law.
3 A robot must protect itself from harm, unless that is in conflict with the first or second laws.

These rules ensure that Asimov's robots are a force for good and that when they take over the administration of human affairs, as they inevitably do, it is as Platonic guardians rather than ruthless autocrats.

Asimov's stories are remarkable for the fact that several of them were written right on the eve of the invention of the computer. From about 1950 onwards, when the computer had become an established and widely publicized fact, science fiction had a sound, recognizable model on which to base its robots; and it is noticeable that, after going through a rather anachronistic 'Michelin Man' phase in the 1950s, movie robots have followed current fashion in the designs of electronic equipment fairly closely. Alternatively, despairing of improving on reality, they have switched back to the idea of androids or humanoids, disguises which are less demanding, in that they only require electronic entrails to be revealed when the robot gets its come-uppance.

But the computer had a far more important consequence than simplifying the life of Hollywood designers; it transformed the robot from an improbable literary fiction into a feasible scientific objective, a good deal further over the technological horizon than the layman might assume for sure, but there nonetheless. Suddenly many of the issues which science fiction had debated stopped being amusing, hypothetical riddles and became serious subjects for study, even for public debate. It no longer seemed totally absurd, for example, to discuss how one might distinguish between a man and a machine, whether a machine could have a mind of its own or whether it might be possible to build a machine which was more intelligent than a man.

The appearance of the computer also crystallized much of the genuine but incoherent unease about machines which had led to the invention of the robot as a literary device 50 years earlier. Man had yielded a good deal of ground to machines over the previous two centuries: they were undeniably stronger than him; they could perform feats of precision which were beyond his capacity; they could detect phenomena that eluded him, like X-rays and light outside the visible spectrum; they could travel faster and in an element, the air, which he could not master; they could communicate instantaneously over vast distances; they could destroy him with appalling efficiency. A common reaction to this growth in the power and scope of machinery had been to redefine the territory which was defensible by mankind on the basis of what appeared to be its natural philosophical frontiers.

The obvious place to draw the line was along the border between mind and body. The physical realm might have to be ceded to the machine, but that left the traditionally human preserves of language, free-will, creativity and abstract thought inviolate – what was being surrendered was no more than the no-man's land which had always been shared with the animal kingdom. It is not too fanciful to say that the sound of the frontier defences going up is echoed in the language. Tasks which are repetitive and undemanding have been labelled 'mechanical'; the 'machinery' (of government, the law or whatever) is that aspect which is routine and automatic; the 'mechanics' of a problem are those parts that can be solved by rote; and when one politician wanted to express contempt for a rival he called him 'a dessicated calculating machine'. In contrast, 'human' has become an adjective which defines qualities of warmth and compassion, even an endearing fallibility, which machines cannot hope to share.

There was a parallel argument which sought to exclude or exempt mankind from the mechanical system to which nature was being inexorably reduced by science. Darwin had undermined traditional, theological fortifications by reducing man to a position on the evolutionary continuum, and determinist arguments had steadily gained ground in philosophy. Most of the advances in science seemed to support the belief that the behaviour of any system, even man himself, could, in theory, be predicted given sufficient information about its constituent parts and their workings. By the middle of the century, those who sought to resist this line of thought and salvage for mankind the dignity of free-will had been driven back on some fairly desperate philosophical nit-picking, such as the argument that the uncertainty principle, which shows that the behaviour of subatomic particles is predictable only in statistical terms, somehow saved the day.

Even psychology had succumbed to the prevailing trend, for in the postwar years behaviourism had become the predominant theory, in the academic world at least. Behaviourists rejected the importance which earlier schools had attached to introspection and the individual's interpretation of his own experience, and classified such self-knowledge as subjective and, therefore, irrelevant and unscientific. They substituted for the vision of man as a free, if often deeply flawed, spirit a model of man as a machine whose performance could only be measured in terms of his responses to external stimuli. The responses were determined by heredity and could, it is true, be modified by conditioning but this was in itself a mechanical, predictable process. To replace Descartes' proud boast, 'I think, therefore I am' with the statement, 'You respond, therefore you are' was scarcely reassuring to mankind's sense of his own dignity. Moreover, if the brain failed to deliver the required responses, and conditioning did not seem an appropriate procedure, contemporary psychiatry was prepared to bludgeon it back into line by chemical or electrical means.

Under these circumstances it is not surprising that those who struggled, rationally or instinctively, to resist this onslaught of machines and mechanical principles felt the advent of an 'electronic brain' to be the final straw. This was not just another probing attack along the disputed frontier but a daring *blitzkrieg* aimed at capturing the citadel of rational thought. The debate as to whether computers can 'think' rages to this day, but the tactics of the battle were clear from the start. It was undeniable that computers could perform mathematical and logical operations, that they had memories, and that they could take decisions on the basis of the instructions with which they were provided. The question was how one interpreted this accomplishment.

Supporters of the computer, and it is noteworthy that many of them were mathematicians and therefore had a natural tendency to see the logical mode of thought that the computer had mastered as the paradigm one, proclaimed that the war was virtually over; all that remained were mopping-up operations, which developing technology would take care of in due course. They buttressed their belief that, in the natural course of progress, the computer would enlarge its scope to include all the functions performed by the human brain with the argument that the precise definition of 'thought' was irrelevant: if a computer *seemed* to be thinking then to all intents and purposes it *was* thinking. This definition of the brain's performance in terms of the appearance it presents to an observer obviously owes much to behaviourism. The most celebrated formulation of this argument is the so-called Turing test, first proposed as a criterion for judging mechanical intelligence by the British mathematician and computer pioneer, Alan Turing, in 1950.

Turing suggested that a human being and a machine be made to respond to questions put by an interrogator. Both human and machine would be charged with persuading the interrogator, who was in a separate room and with whom they communicated only in writing, that they were human. If the interrogator was unable to decide, on the basis of the replies he received, which was which, the machine would have passed the test. Turing urged the validity of his test by comparing it with a similar situation in which one of the respondents would be a man and the other a woman, and both would have to convince their interrogator that they are a woman. The fact that, in the first test, one respondent was in fact mechanical and the other human was as irrelevant to their success as gender was in the second test. This amounts to saying that, if an imitation is sufficiently convincing in practice, then it is pointless to try to distinguish it from the real thing in principle.

One possible counter to this line of reasoning was to argue that, however superficially impressive the computer's performance, it was entirely due to the skill of its designers and the ingenuity of those who devised its programs, and was thus a product of human, not mechanical, intelligence. To meet this point, advocates of the machine's claim to intelligence again drew on the behaviourist model of man. The potential that is built into the computer in the form of its logical circuitry, they suggested, is an exact analogue of the neurological equipment with which man is born, and programs the equivalent of human experience and inherited aptitude. For, in behaviourist terms, the human response to a stimulus is just as much a product of a program or mechanism as that of a computer; to suggest that, because the machine is built by man and its programs planned by man, its performance is somehow inferior in kind is subjective and irrational.

The defenders of the humanist position naturally saw things differently. Though a salient had certainly been driven into territory to which man had hitherto laid claim, the setback could be discounted by reclassifying the ground lost as essentially mechanical.

Abandoning a good deal of intellectual impedimenta en route, the rearguard fled to new positions on the high ground offered by human creativity, imagination and even emotion. From there they issued new challenges – could a computer write poetry? Or compose a symphony? Or fall in love? The best response their opponents could come up with was 'Wait and see'; this could, for the time being, be interpreted as acknowledging that the war had become one of attrition. The supporters of the computer are far from giving up the struggle; researchers in artificial intelligence have continued to snipe away at the outer defences of the humanist position; from time to time they capture an outpost – the success of computers in playing chess is an obvious instance – and occasionally launch a ranging shot at a more ambitious target. But the overall position has not changed as much, perhaps, as they hoped. There is little doubt, however, as to which side is the besieged and which the besiegers.

The relevance of this debate, which became extremely heated in the 1950s, to the robot was twofold. The arrival of the computer at last provided the mental equipment that was necessary if robots were ever to lurch out of the realm of fiction and into reality. As things turned out, however, it took a surprisingly long time, for technical reasons which will be examined later, for the computer 'brain' to find a home in a mechanical 'body', and the real-life robot did not emerge from its womb in the laboratory until the late 1960s. Thus far, although an energetic infant which is already showing its muscle as a competitor in the labour market, it is far from matching up to its fictional counterparts.

In the short term, the advent of the computer had a much more direct effect upon the fictional robot: it gave the idea a plausibility it had hitherto lacked. Previously, it had not been possible to pin the robot down: Capek and others had used it as a symbol of something nasty stirring in the machine shop; Fritz Lang, in *Metropolis*, had employed it to represent a general mechanical malevolence, and even Asimov, for all his effort at realism, had had to postulate an invention, the positronic brain, that had no basis in contemporary science. Most writers had not concerned themselves too much with scientific probability, feeling it sufficient to maintain that somehow, sometime, and somewhere an updated version of Frankenstein's monster would appear and, like some artificial Samson, pull the temple of technology down about mankind's ears.

The computer changed all that and provided a ready-made

'personality' for the robot. It would not be a steel *golem* that wrought physical mayhem but a transistorized bureaucrat which would end the world not with a bang, but a printout. Again, Asimov's uncanny prescience is shown: the final story in *I, Robot* is perhaps the first in which the robot triumphs over man, not by physically destroying or enslaving him but simply by being the ultimate know-all. HAL in *2001: A Space Odyssey* is typical of the new kind of robot: bossy, omniscient, and, ultimately, megalomaniac. It stands for all we dislike and fear about the computer.

The whole history of the fictional robot, from the *golem* to the electronic dictator, has echoes in the legend of the *doppelgänger*. The original *doppelgänger* was a phantasmagoric double, the appearance of which portended the victim's death. There is obviously something extremely eerie about the notion of a twin stepping out of a mirror, as it were, to break the bad news that the time has come; and much of its power rests on the implicit message that the individual is no longer unique and, therefore, no longer 'real' to himself. Transpose this idea, via mechanical man, into electronic man and you have a *doppelgänger* which is doubly terrible. It beats man at the game of thinking, which he had always claimed to be exclusively his own, and, in doing so, it clearly signals, if not physical destruction, at least the metaphysical death of our long-cherished sense of our own uniqueness. The number of the beast is no longer 666 but the computer's binary equivalent.

The expectation, or apprehension, that a mechanical *doppelgänger* with these qualities would eventually materialize has led to much speculation in science fiction and elsewhere as to how we should react to it. If it proved as intelligent as man, what justification could there be for failing to treat it like a man? To treat it as inferior would seem to be a form of racism or slavery, or is there some intrinsic value in flesh and blood which makes it superior to steel and silicon chips and which could warrant a system of apartheid? The problems multiply if the supposition is followed through to its logical conclusion: that a technology that could produce a machine that matched man would soon contrive one that surpassed him. In that case, has man not, by a process of Lamarckian evolution, produced his own successor and, if we hold to the principle that we use to justify our own authoritarian attitudes towards nature, would we not be bound to step aside and make way for the superior being?

None of these problems has yet taken on any practical urgency; even if the Turing test is accepted as a valid criterion, a machine that could pass it with conviction is some way in the future, and the current generation of robots shows no inclination to launch a slaves' revolt: the problems they pose are common to all the machinery of industrial production. The robots of today may be costing jobs but they are not running for office or asking for our daughters' hands in marriage. Are we justified, then, in ignoring the vision as a macabre fantasy of no more practical importance than the prophecies of Nostradamus?

The question can only be partially answered by the science of robotics in its present state of development, but this book will argue that, while the prophets of the robot have been remarkably successful in a number of ways, they have largely failed to foresee the implications of 'a machine modelled on man' in one very important respect.

The assumption of virtually all serious speculation about robots has been that they would be produced by, in one way or another, extrapolating existing machinery (it was of course only with the arrival of the computer that the way ahead could be in any way defined) and would therefore be like known machinery in that they would be determinist and predictable. Like behaviourist man, they might behave in ways which would be superficially inexplicable but understanding would follow if one knew the secrets of their mechanisms and the rules they were following. It was further assumed that, if robots matched or surpassed human abilities, the criteria for judging them would be the behaviourist one, based entirely on the evidence of their output. From this point of view, it would be essentially irrelevant to ask whether a robot was producing the same result because it was working in the same way; all that would be important was that the result was comparable.

The conviction that the utilitarian ends they achieved rather than the engineering means they employed were what would matter about robots was strengthened by the success, in certain areas, of the computer. It is now beyond dispute that computers can store and retrieve information in quantities and at a rate that far excels any human brain, and the same applies to the manipulation of that information according to logical rules. But it is also the case that those who have tried to extend this facility into other areas, such as the use of language, the organization and interpretation of sensory information, or the recognition of patterns, have encountered difficulties which are only partially resolved by the application of larger and larger increments of computing power. A suspicion is growing among some roboticists that, if their machines are to match human performance in a whole variety of fields, it may be necessary to look rather more closely at *how* we ourselves do things as well as *what* we do.

It is certainly true that there are many striking parallels between the computer and the brain, but it is also true that this is the result of what an evolutionist would call convergence rather than any intention to imitate the brain, and it is beginning to look as if the differences may be more significant than the similarities. In short, it may be that the computer's inability to match the brain in some very important respects is due to the fact that it does not work like a brain and, moreover, that simply making it bigger and bigger will never produce a result that satisfies even the behaviourist observer.

Paradoxically enough, the difficulties have arisen not only at the level of creative thought and imagination – the territory so ardently defended by the computer's antagonists – but also at the very mundane level of visual perception and physical coordination, clearly matters that are close to the roboticist's heart. The limitations can be illustrated by what might be called the 'crane driver's conundrum'.

If a crane is picking up a heavy object in one place, say the quayside, and depositing it in another, say a ship's hold, the driver of the crane will have to take account of the fact that the load, suspended from the end of the jib, will behave like a pendulum. It will lag behind the jib when it first starts to traverse and will swing ahead of it when it is halted. A great deal of time would be wasted if the driver had to wait for the oscillation to cease before he lowered the load into the hold, and a skilled driver quickly masters the art of manoeuvring the jib actually to take advantage of the pendulum effect. The skill lies in halting the movement of the jib momentarily before it is actually over the point where the load is to be deposited and then moving it on at exactly the right moment to intercept the load as it swings back and bring it to rest. Moreover, a skilled driver can instinctively adjust to allow for the variables, such as the weight of the load and the length of the cable. The interesting fact is that problems of this kind have proved remarkably resistant to solution by the processes of calculation upon which a computer relies and it has become clear that, even if an orthodox computer were developed to master the 'conundrum', the methods it employed would be quite different from those the crane driver uses.

It is beginning to look as if the attempt to force the computer into roles it is not equipped for will have to be replaced by a different approach based on trying to imitate more closely the workings of the brain as well as its output. Essentially this amounts to abandoning the effort to extract a comparable output from a machine based on a traditional, mechanical model and instead trying to build machines that are based on the biological original: the components will remain the same, but the way in which they are organized will be totally different.

The idea of trying to learn by a direct comparison of mechanical and biological systems is not totally new, though it is only comparatively recently that it found its prophet in the shape of Norbert Wiener, the founder of the science of cybernetics. Wiener defined cybernetics as 'the science of controls in man and machine' and it is, of course, one of the cornerstones of robotics. But its importance for the present argument lies in the fact that it opened up opportunities in medicine, the study of man, as well as in engineering. Wiener, himself, for example, used a variation of the 'crane driver's conundrum' to analyse the malfunction in the human nervous system which causes the disease *ataxia*.

Evidence is beginning to accrue that adopting a similar approach to other aspects of the workings of the brain will result in a better understanding of it as well as enabling us to design machines which can reproduce its workings. Such machines would give robotics a new and important dimension, for the effort to build better robots would become an attempt to understand man, rather than simply to equal or surpass his abilities by other means.

This is not to say that all roboticists will withdraw into the laboratory and try to produce working models of themselves or that all robots will take the form of prototype men. The vast majority of them, no doubt, will work for a living in factory or office and will pose a growing problem as their advantages over human workers become more apparent. We shall have to cope with *robot industrialis* as best we can, taking comfort, perhaps, from the fact that the competition it offers is economic rather than metaphysical – after all, it would lose much of its edge as a competitor if it developed human traits to the extent of requiring coffee breaks or annual holidays.

Robot fundamentalis, though, does not present quite the threat that was foreseen. Rather than being an aggressive interloper from the mechanical sphere, it begins to look more like a willing surrogate for man on which we can perform experiments and test hypotheses which we would find insupportable if the guinea pigs were human. It is not necessary to see such 'clinical robots' as exact replicas, or indecently realistic parodies, of human beings. Instead they would make available to science a new and valuable tool. We are well accustomed to the idea that the use of animals for medical experiments is based on the use of an analogy, that 'man is an animal', and the fact that we find it useful to test drugs on rabbits or the effects of radiation on monkeys does not imply any belief that people and animals are equivalent. Instead, the assumption is that people and animals are very similar in some important respects while remaining quite different in others. The use of a new analogy, that 'man is a robot', need not, therefore, involve diminishing our own sense of the value of being human.

The most important area in which the analogy of man-as-robot could be used is the study of the brain. Again, we do not have to assume a literal equivalence: it would certainly be beyond the reach of any foreseeable technology to reconstruct the electrochemical mechanisms of the natural brain, but it may be within our technological grasp to produce machines which process information in the same way as the brain. It does not take very much imagination to see that a project of this kind could throw new and exciting light on a whole range of issues which have remained tantalizingly resistant to the most adventurous human minds for centuries. Some of them are among the most profound which philosophy has faced: there is the question of determinism, or the importance to us, as individuals, of nurture as opposed to nature (for such a machine would surely have to be exposed to experience and educated just like a human infant); the relationship between mind and body (to the roboticist this is not an abstract riddle but a practical engineering problem) and the nature of 'mind' itself. In a paradoxical reversal of the final scene of

RUR it seems that, rather than having to dismantle robots in order to see how *they* work, we shall try and assemble them to find out how *we* work.

Those who object to the project on the grounds that to probe the innermost secrets of our minds is either impious or imprudent can at least take comfort from the fact that, if and when they confront the *doppelgänger*, it will be recognizable: we will not have created the mechanical equivalent of Frankenstein's monster – we will have re-invented ourselves, or at least that aspect of ourselves which we like to think of as exclusively 'human'.

2. What Is a Robot, AD 1984?

Come listen my men, while I tell you again,
The five unmistakable marks
By which you may know, wheresoever you go,
The warranted, genuine snarks

Lewis Carroll, 'The Hunting of the Snark'

No existing robot looks any more like the robotical fantasies of science fiction than, say, the Mercury space capsules looked like the *USS Enterprise* in *Star Trek*. But, while spaceships may one day match up to our conceptions of what they 'ought' to be like, it is very unlikely that robots will do so. Fictional robots are, as we have seen, designed mainly for dramatic effect; real-life robots, on the other hand, strive, like all other machines, principally for utility.

The pioneers, like Joseph Engelberger of Unimation Inc, who conceived the industrial robot, and nursed it through a difficult childood to see it reach its present healthy and exuberant adolescence, were not motivated by any wish to turn the fantasies of sci-fi into reality, or, indeed, by a consuming interest in man-machine comparisons at a fundamental level. What they saw was that a whole group of technologies had advanced to the point at which it was possible to build machines which could provide a direct and economically advantageous substitute for human labour. Engelberger's own book, *Robotics in Practice*, puts the case with total frankness. It was, he says, the rising cost of labour in the post-war years and the difficulty, in a booming labour market, of finding employees prepared to do unpleasant, boring or repetitive jobs which gave the early robots a toehold in the workplace. Once they had proved themselves, other advantages became obvious: robots do not go on strike or demand pay rises; they will work, uncomplainingly, round the clock for shift after shift; and they make no complaints about the jobs they are asked to do.

In pursuit of efficiency and economies of scale, industry had broken many productive processes down into simple, repetitive stages which demanded little skill on the part of the assembly worker or machine operator, widening the gap between the qualified craftsman who set and maintained the machines and the vast majority of workers who simply served them. It was, as Engelberger confirms, the fact that so many jobs in industry had been almost entirely denuded of skill which made it conceivable that they could be taken over by robots.

The science, or technology, of robotics – and it is worth noting that, just as the word robot was given to the world by a dramatist, Capek, the word robotics was coined by a novelist, Asimov – certainly aims to produce robots which match up to our working definition of 'a machine made in the image of man', but only if the word 'image' is used in a general rather than a literal sense. To try and produce the sort of walking, talking mechanical doll that springs to most people's mind when they hear the word robot would almost certainly be pointless. To start with, we have a more than adequate supply of the original model, and can add to it by methods which are more reliable – and possibly more satisfying – than research in advanced engineering.

Moreover, while there are some aspects of the human model upon which we can certainly improve, strength and durability for example, and others which we may arguably be able to match, intelligence, perhaps, and sensory perception, there are others, like cellular metabolism or the mechanisms of reproduction, which are quite beyond the scope of any known technology. Finally, there are many human characteristics which it would be quite undesirable to reproduce: who, after all, would want a robot that required eight hours of sleep a day and an annual holiday, or which aged, or was prone to disease?

What robotics is concerned with, then, are those human attributes which it seems both useful and feasible to try and imitate; but it is far from clear that there is any advantage in incorporating them all in a single machine, a *robot universalis*. It seems much more likely that, on the principle of 'every robot to his trade', the robots of the future will be specialized and disparate. To be accurate, therefore, it is necessary to qualify our definition so that it reads something like 'a *category of machines* which are, *in one or more respects*, made in the image of man'.

In order to answer the question at the head of this chapter we shall initially concentrate on the sort of industrial robots currently listed in manufacturers' catalogues. This is the best starting point, if only because it was in the industrial setting that the robot first became an identifiable species and the contemporary industrial robot is, therefore, a representative example of the 'state of the art' which will be examined in the following three chapters.

Since the industrial environment has, almost by definition, been a mechanical one for the past century or more, the first question must be: what distinguishes a robot from the general run of machines? There are several different answers to this question, depending upon whom is being asked. An industrialist, concerned with what machines do rather than how they do it, would see the most distinctive features of a robot as its ability to replace human workers more or less on a one-to-one basis and the flexibility which enables it, like a human employee, to perform a variety of different tasks. A production engineer, looking at the way in which machines work, would refine this argument by suggesting that the possession of an articulated limb, analogous to the human arm, is a *sine qua non*. An electronics engineer would, on the other hand, emphasize the control aspects which make robots, unlike much other machinery, autonomous: they must repeat a sequence of operations without the need for human supervision. If they are also required to adapt to different roles, this means that they must have memories of some kind in which to record the steps they are required to go through in order to accomplish a particular job and, in order to meet the criterion of flexibility, the memory must be capable of alteration or amendment.

It is, of course, a function of most machines to replace human labour, and many – sewing machines or electric blenders, for example – take the place of a single worker; some, like mechanical diggers, employ an articulated arm to do so; yet others, such as automatic washing machines, dispense, at least for the duration of a single cycle of operation, with human control. But it is hard to think of a machine, other than a robot, which combines all three attributes, and even harder to think of one which is also adaptable to varying purposes. An industrial robot could thus be described as the only kind of machine which can take over tasks hitherto performed by a human worker, possesses an articulated limb, can memorize and repeat a sequence of operations without supervision, and can be adapted to a variety of purposes. This description fits with the more general definition of 'a machine made in the image of man' to the extent that the only other known 'machine' which matches up to it is man.

Despite the aura of high technology that surrounds it, the sort of role for which the industrial robot was conceived was scarcely glamorous or, in human terms, challenging. By about 1960, when the first industrial robots were appearing on the scene, the tasks that remained unmechanized and unautomated in large scale industry were – if they were not so skilled as to be beyond the capacity of any practicable machine – usually unskilled and mundane: presenting workpieces to machines and removing them again, transferring parts from one conveyor belt to another, packing and stacking finished products. In short, a lot of them boiled down to picking things up in one place and putting them down in another. It was, of course, precisely this kind of work that made many industrial jobs boring, repetitive and unfulfilling and gave the worker the feeling that he was the servant rather than the master of the machines.

The first robots, now often demoted to the status of manipulators, were designed to replace human labour in this kind of niche. What was required was a machine which could consistently grasp an object, lift it, move it to a new position and set it down – over and over and over again. The arms and grippers ('end effectors' in robotic parlance) drew for their design, in so far as they were not derived from commonplace engineering practice, on the comparatively new field of telecherics, what might be called the 'crazy tongs technology' of reproducing, at a distance, the movements of a human hand and arm. The need to handle dangerous materials such as radioactive substances with precision from behind a protective screen had led to rapid advances in telecherics, which had, in turn, also benefited from work in prosthetics, the design of artificial limbs for the maimed or handicapped.

It is perfectly possible to build such manipulators on well-established engineering models; indeed, they are commonplace in one form or another. Most printing machines, for instance, employ sets of grippers in pairs to pick up each sheet of paper in turn and insert it into the press. But, clearly, manipulators of this kind are highly specialized, designed to work on one particular material or component in one particular context. Moreover, they operate on traditional mechanical lines; the movements of the grippers are dictated by systems of cranks, levers or cogs and the only way of altering those movements is, in effect, to redesign the machine.

An arm with three or more articulations provides the means of moving a gripper into an infinite number of positions within a given space and, especially if the 'wrist' joint can flex in two or three planes, of orientating it in a wide range of different attitudes; and the power to operate the articulated joints is available from well-tried electrical, pneumatic or hydraulic devices. The problem comes with the need to store and, when necessary, change, the 'instructions' which will tell the joints which way to flex, by how much, and in what sequence. The methods that were used in the first generation of manipulators relied upon electromechanical switches (switches which close when a current is applied to an electromagnet and open when it is switched off, similar to the solenoid which operates a car's starter motor when the ignition key is turned) triggered in sequence by a device rather like the automatic timer on a central heating system. Flexibility was obtained by linking the switches to the controlling 'clock' via a 'pinboard' which resembled an old-fashioned manual telephone switchboard.

The snags in such devices were twofold. Firstly, the business of resetting the pinboard to 'program' a new operation was tedious and lengthy, often involving days of work; in a sense it was still a question of redesigning the machine or, in computer terms, modifying the hardware rather than the software. Secondly, the degree of control that could be obtained by a series of on/off switches was very crude; the instructions they gave, in the form of bursts of energy delivered to a particular motor or valve, were elementary: 'move' or 'stop moving'. In practice the instruction to 'stop moving' could seldom be made sufficiently accurate and mechanical stops and limit switches (switches which were operated by the arm coming into contact with them) were required in order to position the grippers with any accuracy. The need to rearrange these stops and switches added to the time taken to reprogram the manipulator for a new task.

The computer provided a means of radically improving the manipulator's skills, to the point where it could properly be called a robot. Not only could it store the whole sequence of instruction in electronic form, it could also receive and act upon information provided by sensors on the arm itself. We shall be looking at the whole, crucial question of control, feedback and the contribution of cybernetics in a later chapter; for the present it is sufficient to note that accuracy can be improved dramatically if the instructions that the arm receives take a form such as: 'move and continually report your position until I tell you to stop moving'. Programs which could be recorded on tape or disc and, if appropriate, stored and reused, offered an enormous improvement in flexibility. This was enhanced by what is called the playback technique. Instead of the operator working out a sequence of instructions in the abstract, a playback robot is programmed to allow it to record its own instructions as the operator leads it through the required movements, either by physically moving the arm or by means of a remote control panel which he holds in his hand.

A robot with computer control could aspire to tasks quite a bit more ambitious than simply picking things up and putting them down with its grippers. It could, for example, be equipped with specialized end effectors such as electromagnets or suction cups that would enable it to handle things like sheet metal or glass which were too large or unwieldy to be grasped by mechanical claws. If a welding tool replaced the gripper, it could carry out work such as spot welding on a car assembly line, where the premium is on speed rather than skill or extreme accuracy. If it was given other end effectors, such as paint guns or ladles, it could do work which was hazardous or simply impossible for humans to carry out: spraying car bodies with toxic paints or loading furnaces or moulds with molten metal. It was also preferable if it was a robot arm rather than a human one which undertook dangerous tasks such as placing blanks in a press bed beneath a hammer weighing several hundred tons and delivering ten or more strokes each minute, especially as the robot could communicate electronically with the press in a way not open to human operators.

In fact, the range of jobs that robots could tackle widened rapidly, but until the mid-1970s industry was curiously slow to take advantage of their growing skills. The reason for this lies in the way in which the computer developed.

For the first twenty or so years of its life, until the late 1960s, the computer was afflicted with gigantism; the forces which influenced its development, technical progress, and the commercial interests of manufacturers and purchasers, all pointed towards bigger and bigger machines. The hardware, like transistors, magnetic memory cores, etc, that went to make up a computer was still very expensive and it seemed likely that it would continue to be so, effectively limiting computers to a market made up of large organizations. The strategy adopted by the manufacturers, and usually credited to a market analyst at IBM called Herb Grosch, was to accept this but to maximize the market by making computers as large and flexible as possible. Instead of persuading customers that they needed a computer to do this particular task or store that special kind of information, manufacturers offered machines that could fill a whole variety of roles, thus offering the purchaser savings or increased efficiency across as wide a spectrum as possible.

The giant 'mainframe' machines of the Grosch era could cope, via remote terminals, with a whole range of tasks simultaneously (simultaneous is not a strictly accurate description, but that is how it appeared to the user). This not only meant that a large business could, for example, use a single machine to run its payroll, control its inventory, do its book-keeping, produce invoices, and control credit, but that smaller companies, which could not justify actually buying one of the giant machines, could gain some of the same advantages through access to computer bureaux and 'time-sharing' arrangements.

The best that an industrial robot could look for in this situation was the chance to 'plug in' to one of the big mainframe machines and be granted some small fraction of its 'attention'. The arrangement had the effect of retarding the robot's development for two reasons. First, the amount of brain power (strictly speaking, the amount of processing time) which a large computer could spare for robot operations was limited, especially when it came to the business of changing programs which demanded more of its capacity than simply running them. Second, the possession of computers, and hence the availability of sockets into which the robots could plug, was limited to large organizations that, in turn, tended to be interested in robots as a *lumpenproletariat*, useful for only the simplest and least skilled work. This is because it is in the nature of big businesses to concentrate on long production runs of standard products and leave more specialized work to the smaller fry who could not afford computers and, in any case, needed skills and a degree of adaptability which existing robots lacked and which were not required in the context of large scale mass production.

Most of the largest and best known robot installations that exist today, like the Fiat 'dark factories' in Italy, are products of the 'Grosch' era. The robots are used as essentially unskilled labour to fill the awkward gaps that automation has been unable to bridge in other ways. They are certainly impressive, even uncanny, but they are nonetheless mere dependants of a computer technology that is as obsolete as the dinosaurs.

The umbilicus tying the robot to the big computer was severed in the early 1970s by the silicon chip. The coming of the chip not only shrunk the physical dimensions of the computer, it also cut its cost even more dramatically. From being an apparently insurmountable obstacle to the spread of computer use, the costs of hardware became, over a decade, an almost negligible factor. The first consequence for industrial robots was that, instead of having to compete for the attention of a big centralized brain, they acquired a brain of their own in the form of a minicomputer 'dedicated', as the jargon has it, to controlling them. The mini was, in turn, quickly replaced by the microprocessor, essentially a complete computer on a single chip. In scarcely a decade, the robot's brain had shrunk from a vast, remote roomful of gadgetry costing millions of dollars and requiring air conditioning and the services of white-coated acolytes to something the size of a shoebox, costing a few thousand dollars at most, and easily bolted on to the robot anatomy in any convenient position.

In the mid-1980s, it is fair to say, the industrial robot has available to it as much brain power as it can use, though it is important to qualify this statement by noting that the brain power is of a particular kind which imposes rather severe limitations. The microprocessor, however small and cheap it may be, is still a computer and, as we shall argue later in this book, the computer may not be capable of filling the role of all-purpose robot brain. But leaving this major point aside for the moment, it is possible to discern a number of ways in which the industrial robot is likely to use its newly found mental resources in order to widen its repertoire and sharpen its skills.

The following chapter looks in some detail at the engineering of robot anatomy, but one general trend is clear. Many robots are already breaking away from the anthropomorphic model they inherited, via telecherics and prosthetics, from the human original. This is entirely logical; much of the need for machines and tools of all kinds arises out of the fact that engineering can obtain power, precision, and other advantages that biological systems cannot match. As Buckminster Fuller pointed out, technology has generally advanced by replacing devices that had a base in nature, the crank for example, with ones that were entirely artificial in conception, like the turbine, and it is reasonable to suppose that the robot will do likewise. This has happened almost from the start in some instances. For example, many robot arms take the form of telescopic rams which extend or contract to lengthen or shorten their reach, whereas nature, lacking this mechanism (the only human organ to operate on even remotely similar principles is notoriously incapable of continuous, repeated operation and is, moreover, confined to the male sex) has to make do with combinations of flexible joints which are inherently more complicated and difficult to control. In particular, it seems likely that the need to simulate the hand will largely disappear: hands in industry, after all, are commonly used to hold tools, and the robot can short-circuit the process by plugging tools directly into its 'wrist'.

Robot anatomy will also be influenced by another development that comes in the wake of the microprocessor. It is logical to take the process of devolving the powers of control which began by freeing the robot from its reliance on the multi-purpose computer even further by equipping it with two or more microprocessors, each responsible for a different aspect of its work.

The advantages of such an arrangement are similar to those that can be obtained by breaking any complicated operation down into its component parts. The job of stuffing, sealing and stamping a large number of envelopes will, for instance, be simplified if one person folds the enclosure, another inserts it into the envelope, a third seals it and a fourth applies the stamp, for each worker can develop a simple repetitive cycle. In the same way, a microprocessor which can concentrate, say, on controlling the wrist articulations of a robot can be simpler, cheaper and more readily reprogrammable than one which has to cope with all the robot's operations simultaneously. There are also gains to be made in the speed at which a robot can work. A single processor, though it may handle thousands, or even millions, of calculations a second, can only do them one at a time; the possession of several processors will allow a robot to 'think' about several different things at once. There will obviously have to be intercommunication among them, but this can be limited to essentials.

Interestingly, in following this line of development, robots will be modelling themselves on some relatively primitive creatures like ants in which the control of limbs is devolved to a secondary nerve centre in the spinal cord. Parallel researches in computer science are also seeking to speed up calculations by distributing them among a number of processors rather than relying, as the 'traditional' computer does, upon a single, central processor for all logical operations. In so far as it can be compared to a computer, the human brain can also be said to employ 'distributed processing' – it seems, for instance, that when we 'go for a walk to think something out', the walking and the thinking are handled by different parts of the brain.

However they are achieved, the benefits of distributed processing are likely to become particularly evident as robots gain a more elaborate set of senses. The information that contemporary industrial robots contain is usually limited to their own anatomy; they will 'know', for example, that their elbow joints are bent at an angle of 60 degrees or that their wrists are rotated clockwise 30 degrees from the horizontal. In medical terminology most robots have only proprioreceptive senses. Some, equipped with a rudimentary sense of touch by means of compressible pads or contact switches, can judge whether they are actually holding something in their grippers, but in general the only knowledge they have of other objects and their environment is that relayed by other machines. A conveyor belt may, for example, be equipped with a switch or photo-electric cell which tells a robot that an object is in position and awaiting collection; a press will be able to inform its robot servant that it is locked in the open position and ready for another workpiece to be placed in the pressbed; or one robot may inform its fellow that it has completed an operation and that a component is now ready for the next stage in a process.

In some respects this ability of a robot to communicate 'on the same wavelength' as other machinery enables it to cooperate in a way in which human workers cannot – an instantaneous electronic 'job finished' message from a machine tool, for instance, can be acted on by a robot more speedily and reliably than a human operator would be able to react to the evidence of his eyes or ears – and, of course, there is no danger of the robot's mind wandering to last night's movie or this evening's dinner. And, even if the worst occurs, a smashed end effector is infinitely preferable to a mangled hand.

But it is also obvious that a robot which receives only this sort of information is severely limited. It relies on everything being entirely predictable: each object it is to pick up being presented in the same place and in the same attitude or, if it is carrying out something like a welding operation, a jig which holds each workpiece in exactly the same position. This can be a serious handicap in even quite elementary tasks. In order to train a robot to stack boxes on a pallet, for example, the operator will have to treat the placing of each box as a separate operation, first programming the robot to put the first box in the far left hand corner, then the second box one space to the right . . . the tenth box on top of the first, the eleventh on top of the second, etc. No one will wish to embark on this sort of process if the job is a transitory requirement; it only makes sense if identical boxes are going to have to be stacked on identical pallets week in and week out over a long period of time.

A robot which could see what it was doing would clearly have enormous advantages. Providing it had the ability to interpret what it saw and take the appropriate action, in other words if it could recognize even relatively simple patterns and react to them with intelligence, it would not be necessary to 'nursemaid' it to anything like the same extent as an unsighted robot. Objects could be delivered to it higgledy-piggledy; it could even be expected to distinguish one object from another and sort them into categories; given some general rules about how pallets should be stacked, it might learn to cope with a variety of boxes or pallets of differing sizes.

The production of a means of vision for robots is currently the top priority in research, so great are the advantages perceived to be. There are a number of systems at quite advanced stages of development employing various techniques. The differing approaches divide, however, into two distinct categories: the preprogrammed and the adaptive. In order to understand the distinction it is first necessary to grasp where the real difficulties lie. There is nothing problematical about providing a robot with a picture of its surroundings, or in focusing in on a particular object, magnifying it or rendering it in glorious Technicolor. Television cameras can do all that, and have the advantage of reducing the information to digital form which a computer can accept. The difficulty comes when one tries to enable the computer to 'understand' or recognize what it sees.

Preprogrammed techniques attempt to solve this problem by getting the computer to analyse a visual image in a variety of ways: finding the corners or edges of objects, their positions relative to each other, and so forth. Chapter Nine describes the main methods which have been developed and examines the reasons for their comparative lack of success. The problems which arise are all, in one way or another, due to the enormous amount of information which the computer has to absorb and analyse before it can reach even a fairly elementary conclusion. It is perfectly possible to program such a system to distinguish unfailingly between, say, nuts and bolts – provided that the bolts are always presented sideways on, so that they are recognizably long and thin in contrast to the nuts which are more or less circular from any angle. Unfortunately, in the real world, bolts may turn up in any orientation and, end on, will look a lot like nuts. The more complex the object, and the finer the distinctions to be made, the more elaborate the program required.

Adaptive systems work on quite different principles which, again, are examined in depth in later chapters; for the present it will be sufficient to say that adaptive systems, rather than programming the brain to analyse an image according to set rules, aim at translating images into generalized electronic patterns which the machine can store and recognize.

Preprogrammed systems require preprogramming in that the image that is to be recognized has to be broken down into data, rather as a surveyor records topography in a series of elevations and coordinates. Adaptive systems are adaptive in that the hardware translates any image into a pattern dictated by its internal construction, and the system can therefore quickly adapt itself to recognizing a new image when and as required. This book will argue that the long-term future lies with adaptive systems (and not only for vision) but in the short term both ideas will no doubt be developed and used.

Recognizing what is seen will widen the robot's range only marginally. It will certainly enable it to sort nuts from bolts, but if it is to be able to fit them together it will also require 'intelligence'. Again, the subject of artificial intelligence is dealt with in later chapters, but as far as the industrial robot is concerned it must be said that the prognosis is not optimistic for the immediate future. The researches which seem most applicable, those concerned with devising programs which enable machines to recognize and manipulate shapes or objects, have got no further than the stage of a one year old baby struggling with a set of simple building blocks, and to achieve the level of success that most mothers would expect of their infants has involved the use of highly complex programs and disproportionate amounts of computing power. There are, moreover, no realistic ideas as to how even these modest accomplishments might be utilized in the workplace. Yet progress in this direction will be necessary if robots are ever to advance from what is, in human terms, unskilled work to even semi-skilled employment, such as the assembly of intricate components.

How far, it might be asked in conclusion, is the industrial robot of the present and the immediate future matching up to the roboticist's aim of a 'machine made in the image of man'? This question can be looked at from two points of view. In so far as industrial man is viewed as a worker who has acquired skills, the robot has made little progress, partly because much human ingenuity had already gone into inventing specialized machines which could eliminate the need for some skills and partly because skilled work demands intelligence, even common sense, to a degree which has so far eluded the robot. It can, however, be said that the robot has shown great promise as a user of tools, though it has to be added that a robot is good at wielding a paint gun or a welding gun in the same sense as a machine tool is 'good' at using a router or a drill – because the tool is most effectively employed in a way that is mechanical and repetitive rather than inventive.

Adopting a different viewpoint, which sees industrial man as having been reduced to a servant of machines, there is no doubt that the robot has shown enormous potential as a general, unskilled worker. But the fact that it can compete so ferociously, in economic terms, in this field is, it might reasonably be said, a comment on the degree to which large scale mass production devalues and underutilizes human talents. The robot's progress to date might be summed up by remarking that the jobs at which it shows most promise are those which a human worker might describe as being 'so simple a machine could do them'.

This is not to dismiss the robot as unimportant to industry and unthreatening to workers, or to ridicule it as simply another kind of machine which is presumptuously laying claim to a new status. It is simply to say that, as things stand to date, industry has found particular aspects of the robot valuable while largely ignoring others and to suggest that in the immediate future the things that industry will find most useful in robotics are those which distinguish robots from human beings rather than those which link them. Industry has every incentive to encourage robots that can take their place on the production line and work round the clock; it has less interest in robots that demand a seat in the boardroom.

3. The Engineering of Muscle and Bone

The back bone's connected to the shoulder bone
The shoulder bone's connected to the neck bone
The neck bone's connected to the head bone
Oh, hear the word of the Lord

Traditional

In view of the fact that the industrial robot arose out of the need for a machine that could provide a direct, more efficient, or more economic substitute for a human worker, it is not surprising that the anatomy of most present-day robots is distinctly anthropomorphic. A machine designed to fill a niche previously occupied by a person will have, to some extent, to be modelled on its predecessor – it will, for example, cause undue disruption to the industrial environment if it occupies vastly more space or operates in a totally different way. This anthropomorphism may, however, be no more than a passing phase; as robots and automation advance in parallel, the need for robots to be broadly interchangeable with people is likely to be of diminishing importance.

For the moment, however, robotics is set in an anthropomorphic mould to the extent that it has taken one part of the human anatomy, the arm, and, like some obsessional bodybuilder, concentrated on developing it to the exclusion of all other aspects of the physique. Rather as Edwardian caricaturists drew giant heads on incongruously tiny bodies in order to concentrate on the features which were significant to their purposes, roboticists define their products in terms of the bit that matters, and the kind of arm a robot has largely determines the kind of robot it is.

Other than as an anchorage for its 'shoulder', a robot has no need of a trunk; a single electric cable will serve the purposes for which the human body requires a torso – respiration, digestion and excretion – and most contemporary robots are firmly anchored to the ground and have, as yet, no need for legs or an equivalent means of mobility. As for the head, it may be at the end of another cable, or just bolted on in any convenient position where it will be out of the way. This may change as vision systems and other senses become available, but even then it is likely that sensors will be attached to the arm the better to monitor its actions. Some robots indeed are already equipped with a rudimentary eye on the 'palm' of their end effector. Like some baroque spectre, the robot of the future may haunt the factory with its eyes (if not its head) tucked beneath its arm.

Viewed as a piece of engineering, the arm is a device for placing the hand anywhere within a given volume of space. The human arm achieves this feat by having three degrees of freedom, two at the shoulder joint and one at the elbow. The shoulder allows us to move the arm in a horizontal arc of about 180 degrees across the chest (a kind of swivelling version of the fascist salute) and in a vertical arc (the movement marchers use as they swing their arms). Combined, these two degrees of freedom will position the hand anywhere on the circumference of its range. The in and out movement required to reach intermediate positions is provided by the flexing of the elbow.

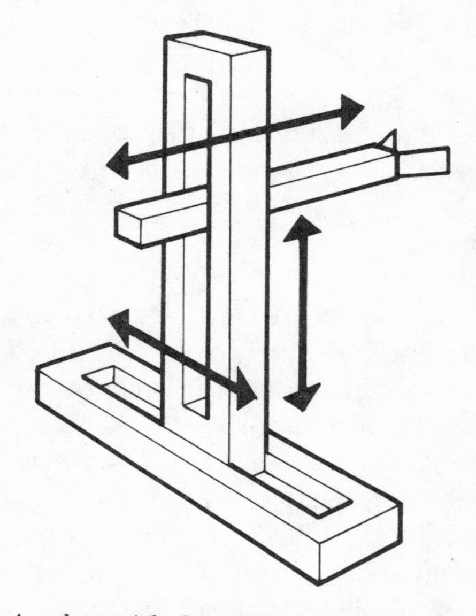

A robot with Cartesian coordinates

All robot arms have three equivalent degrees of freedom, though they are not necessarily engineered in the same way. The simplest system is movement along the classic Cartesian coordinates: up and down, side to side and in and out. A robot using a system of this kind will have great rigidity and be comparatively easy to control, especially since a movement along one coordinate can never result in a displacement along another or alter the orientation of the 'hand' or end effector. Almost the same can be said of a robot which moves on cylindrical coordinates, replacing the side to side movement with a rotational one: this obviously gives greater flexibility and allows the arm to cover a larger area at the cost of only slightly more elaborate engineering.

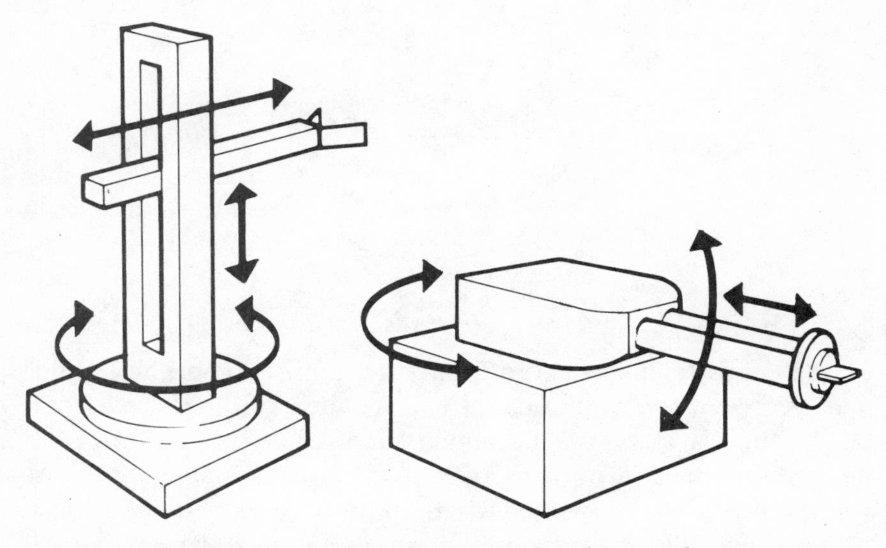

A robot with cylindrical coordinates *A robot with polar coordinates*

A robot operating on polar coordinates has a much more anthropomorphic feel to it. Essentially, it follows the example of the human shoulder and achieves up and down movement by tilting, rather like a naval gun in its turret, but replaces the elbow with a rigid, telescopic joint. It has obvious advantages when it comes to picking things up and moving them, since raising and lowering the wrist is achieved by small, economical movements at the shoulder rather than by raising and lowering the whole arm; the latter arrangement puts the robot in the position of a human who has to bend at the knees to pick something up from a table. Control, on the other hand, becomes more complex, since the tilting movement used to raise or lower the end effector will affect its alignment in the horizontal plane and must be compensated for at the wrist.

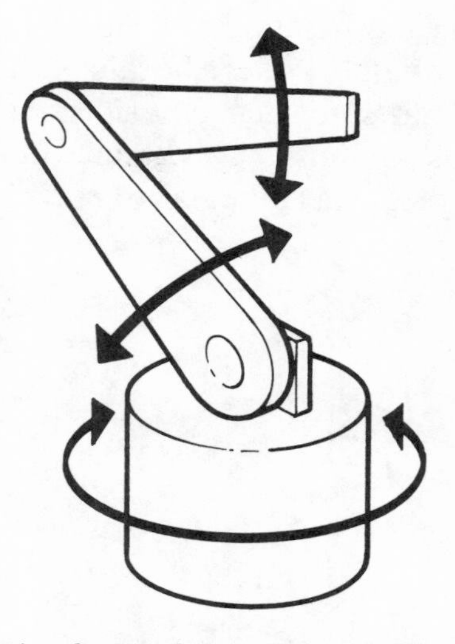

A robot with revolute coordinates

The fourth, most elaborate and most anthropomorphic, system of arm geometry is that based on revolute coordinates. Here the mechanical arm follows the human example in using an elbow articulation, though the normal practice is to turn the human model upside down so the elbow is inverted, like that of a freestyle swimmer or an anglepoise table lamp. An arm using this design can clearly reach over or around obstacles or into containers and achieve positions ('scratching its armpit', for example) in a way the others cannot. Equally clearly, it lacks the rigidity of simpler arms and controlling it raises more complex problems since any movement of the elbow has to coincide with a movement at the shoulder if it is not to cause an up and down displacement as well as an in and out one. There will also have to be movement at the wrist if the end effector is to remain in the same alignment.

Putting the hand in the right place is, of course, only half the battle. As anyone who has ever had his wrist in plaster will know, orientating the hand is equally important. To achieve a variety of orientations, the human arm has a further three degrees of freedom: the up and down, flapping motion of the wrist, a smaller side to side motion and a rotational movement which involves the forearm turning about its own axis. Most robots follow the human model, with the difference that the rotational movement takes place at the extremity of the wrist, so that a full 360 degree rotation can be made

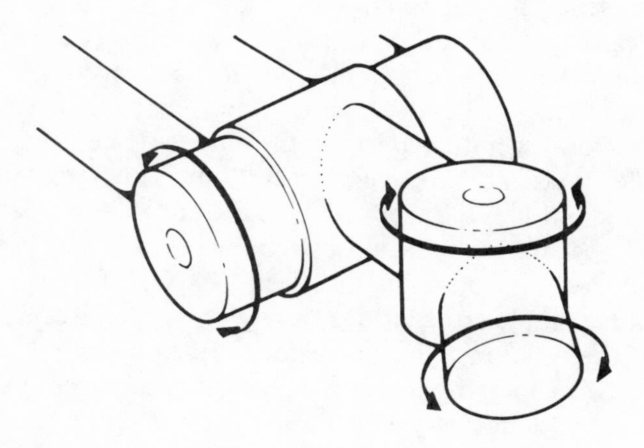

Typical wrist articulations

without affecting the alignment of the other joints. This frees the robot from the need to relax its grip and take a new hold as the hand must do every half turn when using something like a corkscrew or a screwdriver.

Some robots, however, achieve greater simplicity by sacrificing one or more of the degrees of freedom at the wrist, for if use is made of larger, less economical movements higher up the arm, the end effector can still, in theory, achieve all orientations with only five of the six articulations; and in many applications a full set of wrist movements is not required. A robot which can rely on objects being presented in a consistent position, for example, and which has only to pick them up and move them about, may need scarcely any movement at the wrist at all. But most roboticists will look with suspicion on machines that lack all six degrees of freedom, feeling almost instinctively that sooner or later the simplicity will prove a handicap.

However great the superficial resemblance between a fully articulated robot arm and the human limb, great differences remain. The joints in the skeleton all work on the ball and socket principle. This allows great flexibility; the shoulder, for example, achieves two degrees of freedom with a single joint and can thus flex on an infinite variety of diagonals between the horizontal and the vertical planes rather than relying, as the mechanical arm does, on the simultaneous movement of two joints. The ball and socket, however, relies on muscular tension to hold it in place and can be dislocated relatively easily, and it moves as a result of muscles flexing and relaxing symmetrically in extremely complex patterns.

These are characteristics the engineer finds very difficult to reproduce or control, for both depend on the muscle tissue being elastic and elastic materials are very tricky and unpredictable to handle – even our biological equivalents are not totally reliable, or we would all be able to do the most strenuous exercises without any need for practice or 'loosening up'. Robotics has, therefore, been content to use simple articulations flexing in one plane only, or, in the case of the rotation which gives side to side movement, a turntable such as is commonplace in a dozen other applications. This simplicity in the design of joints means that it is not necessary, in every case, to use mechanical muscles in the same way as natural muscles are used, to apply leverage to the skeletal structure; instead, rotational force can be applied directly to the joints via a gearbox, an arrangement which permits greater precision, especially when the force is provided in the form of torque (twisting power) from a motor.

Allotting a separate articulation, rotation or telescopic movement to each degree of freedom means that a separate muscle, or drive, also has to be provided and a robot with six degrees of freedom will require six drives, each with its own source of energy and its own link to the controlling 'brain'. It also means that, unless the arm is always to travel from, say, the top right hand corner of a square to the bottom left hand corner by going first down and then across, anything but the most elementary movement is going to require that at least two drives work in conjunction with one another. Thus, although the control mechanism defines every position in terms of a kind of map reference, the map is three dimensional and the straight line joining any two points in space is the sum of two or three lines or arcs on different coordinates.

If the arm is to be capable of more than simply moving from point A to point B and is to have what is called 'dynamic control', which will prescribe the path it is to travel in order to avoid obstacles, vary its acceleration and deceleration in order to give smooth and fluent movement and take account of the different weights it may be carrying, the speed and torque of the drives will also have to be variable. A robot muscle, therefore, is far more than a simple source

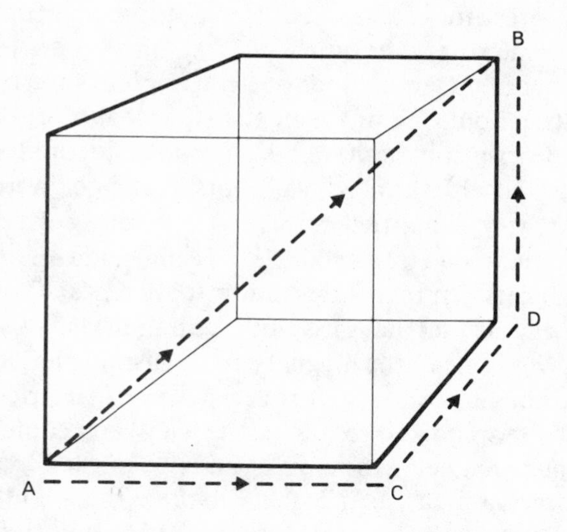

Although the movement of the end effector from corner A to corner B can be represented by a straight line, it can only be achieved by a combination of movements involving all three of the arm's degrees of freedom and will take much longer to accomplish if the end effector must follow the path: corner A to corner C to corner D to corner B

of power acting on a level or rotating an axle; it must push, pull or turn the right way, at the right time, with the appropriate energy and, above all, to a precisely accurate extent. If its muscles are not susceptible to strict discipline, a robot arm will behave like an untrained boxer who loses his temper and flails wildly in all directions.

Robot drives are, with very few exceptions, either electrical or hydraulic. Cables and hoses are both conveniently flexible conduits for energy which can be led to any part of the robot anatomy without getting in the way of its movement. Hydraulics offer a rather cheap technology and are the only choice for robots needing a great deal of strength, since large electric motors are too heavy and cumbersome for incorporation into an arm; but hydraulics are also less precise in operation and mate less easily with a robot's electronic controls. Hydraulic muscles can take the form of rams acting directly on the skeletal limbs – the kind of obvious external muscles that can be seen on mechanical excavators – and rams are the natural choice for a telescopic arm which extends and retracts. But hydraulic pressure can also be used to drive a variety of motors which provide rotary motion.

A number of different kinds of electric motors are used to provide robot muscle, but one of them, the stepper motor, deserves more detailed description because its ingenious method of operation meshes so well with the needs of robots. A stepper motor can be described, in very simplified terms, as a rotating magnetic spindle which is surrounded by a cylindrical casing made up of a number of electromagnets, up to a thousand in some cases. The secret of the design lies in the fact that the spindle remains stable when its magnet is aligned with one of the magnets in the casing, and, if energy is delivered to the motor in the form of pulses of electricity, one pulse will cause the spindle to rotate by one 'step' and come to rest in line with the next magnet in the casing. Thus, to introduce a concept that will become familiar in the next chapter, the spindle has only a limited number of 'states' (equal to the number of magnets surrounding it) in which it can come to rest, and the input of a measured amount of energy will produce an equivalent movement – no more and no less. Since counting is one of the computer's fortes, it is very easy for a robot brain to organize movements which take place in response to a measured number of pulses of electricity, and it is far easier to achieve precision by use of a stepper motor than by starting and stopping the flow of hydraulic fluid through a valve or the supply of current to a more orthodox motor.

When our attention moves from the arm to the hand, the robot's anthropomorphic claims begin to look exaggerated. Mechanical arms may not work exactly like human ones, but they can do as well in many circumstances and considerably better in some – not only is it easy to make them longer and stronger, it is also possible to design variations on the original theme which can move in ways and take up positions that elude even a contortionist. But when we look at the robot's end effector we immediately drop any idea of human comparisons. If we think of any animal at all it is of an unusually sprightly lobster, for (leaving aside special purpose effectors for the moment) a robot's hand is little more than a pair of pincers, and the idea of it matching human accomplishments like playing the piano or dealing a deck of cards seems absurd. The crudity of robot hands is in part due to the fact that in most applications where a robot is required to grasp things a more elaborate hand would be an indulgence. There is, after all, little point in equipping a robot with a skilful hand capable of manipulating complex shapes if it cannot see what it is doing. As the prospect of robot vision looms larger on the horizon, hands are being developed to take advantage of it: a hand with two fingers and an opposed thumb would enable robots, in principle, to grip (though not to manipulate) any object, and a team at Cranfield Institute of Technology in Britain has produced the prototype of a hand with the full five digits. But, given the difficulties of miniaturizing components and achieving dexterity and sensitivity, it will be a long time before we hear a robot virtuoso on the piano or meet a robot card sharp.

A rather more subtle point is that it is doubtful how far most robots are likely to need hands at all. Skilled industrial hands are most usually found operating controls or holding tools. The robot can handle the former function much more easily by direct, electronic communication, and instead of holding tools in its end effector, it simply substitutes them for it. This trend can be seen even when the robot is doing the menial work of picking things up and moving them about. It is much easier, for instance, to handle sheets of glass or plastic by holding them with pneumatic suction cups, or sheets of metal by means of electromagnets, rather than by gripping them in more orthodox fashion; not only is the grip firmer, but grasping and releasing is a matter of switching pressure or current on or off. If it is necessary to pick up a length of pipe, why not use a hand with a giant span that can insert a finger from either end? Or again, instead of giving a robot two arms to clasp a bulky carton, why not simply provide it with one bigger hand?

Taking this process a step further, some robots are using two hands on a single wrist. This has great advantages when, for example, a robot is required to unload one component from a machine tool and replace it with a new one: instead of reaching in, picking up one part, taking it out and then inserting the arm again with the new part, it grips the finished component, instructs the machine to open its chuck or jaws, rotates its wrist through 180 degrees and the new component is automatically in position.

A logical next stage from the specialist gripper is the end effector which is itself a tool. The first two instances of the 'plug-in tool' to be installed on robots, and still the commonest, are the spot welding gun and the paint spray gun. But many other tools, from simple ladles for handling molten metal to socket spanners and cutting heads, have been adapted for robot use. There is in principle no tool which a man can use, and several he cannot, which a robot cannot manipulate with some adaptations. Already systems are in operation which allow robots to place one tool in a holder, disengage their wrists, and plug into another tool. The main factor delaying progress in this direction is that many tools require the use of visual, or other, senses if they are to be used with anything more than rudimentary effect. It is possible to get round this problem in some cases by providing guides which ensure, say, that a spanner finds the head of a nut, but this is obviously a second best solution and cannot be used in cases where real judgement and skill are needed.

The robot's real problems are not what kinds of hands it should have, or what it should do with them, but how to see what they are doing and how to move them about with more accuracy and greater speed. Later chapters will look in detail at control and sensory problems, which are emerging as the central preoccupations of robotics, but it is worth examining at this point the implications which the engineering of muscle and bone holds for those more complex issues. Because a robot arm moves about within its sphere of action without any mechanical constraints keeping it on the straight and narrow, as a cylinder confines a piston or a bearing restrains a crank, controlling it has as much to do with stopping it moving as with ensuring it moves in the right way. It must, if it is not to knock itself and its surroundings to pieces, accelerate and decelerate smoothly and arrive at its destination with the minimum of impetus; otherwise it will treat every workpiece like a host who gives his guests a vigorous right to the belly when he intends a polite handshake.

For a robot of any size, the inertia or impetus of its own weight, let alone that of the object it is carrying, presents a considerable problem. A robot arm, with motors and gearboxes at wrist and elbow, is no lightweight. Try manoeuvring your own arm with a weight of only a few pounds held in the hand and it will become clear (leaving aside the question of sheer strength, which is of secondary importance) that moving quickly from one position to another with any speed requires a lot of practice: the skill required is akin to that of the Japanese swordsmen who can cut a cloth held on an outstretched hand without severing the skin beneath. Yet, if robots are to speed up their operations to a rate at which they become competitive with human workers at light assembly tasks and suchlike, they have to move faster and with great precision.

Part of the problem is that neither electric nor hydraulic motors are nearly such lightweight and efficient converters of energy into motion as muscles, and nor do rigidly hinged metal arms have the built-in elasticity of flesh and bone. One of the most striking impressions one gains from watching robots at work is how often hundreds of pounds of steel are being moved about in order to accomplish some relatively trivial effect. As the population of robots grows, the amounts of energy wasted in this way will become a serious consideration. A solution currently proposed, and being investigated at MIT, is to try and build arms of lightweight, flexible materials which will, literally, whip from position to position. A flexible arm will raise horrendous problems of control, posing difficulties very similar to the crane driving conundrum examined in Chapter One, but building some elasticity into the arm, or its joints, is the only possible way of reducing the time taken for a typical arm movement from the period of several seconds required by current robots to the fractions of a second taken by the human arm. Trying to make machines engineered on present lines move at that sort of rate results in a total breakdown of control (the arm overshoots or undershoots its objective dramatically) and causes the robot literally to rattle itself to destruction.

Another line of research is examining ways of enabling mechanical muscles to operate articulations in a fashion that is closer to the natural model. The idea is that of replacing a single 'muscle' pushing or pulling a joint which hinges in one plane with a number of separate rods arranged to tilt the joint, the rods being connected to the perimeter of the cross section of the joint so that it tilts as the rods on one side extend and those on the other retract. This would have the advantage of allowing designers to combine two degrees of freedom in a single joint, just as the human ball and socket system does, but, again, the control problems are substantial, and the need to provide a separate source of power for each rod is likely to limit the design to small robots used for intricate work rather than large ones which need brute strength.

Current trends in the use of industrial robots are diverging in two directions: towards a closer, more exact equivalence between man and machine, and away from any form of anthropomorphism. The

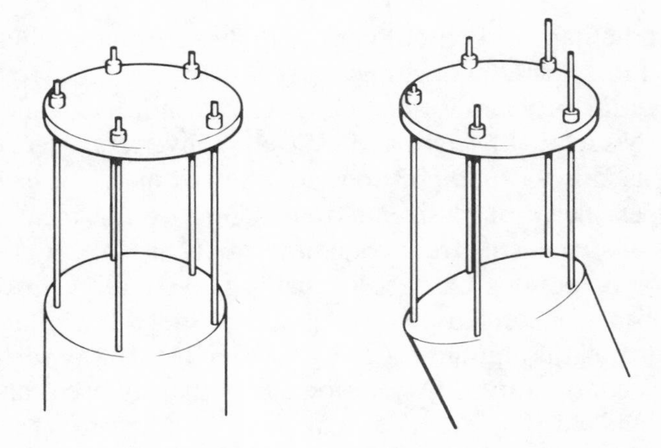

A muscular system closer to the biological model; by coordinating the movements of the rods through the drives on the upper part of the joint the lower section of the limb can be provided with two degrees of freedom

first tendency arises out of the fact that, as robots become more sophisticated, the number of roles in which they can substitute for people is increasing and with it the need to make them more fully interchangeable with humans. This has led to a number of designs for machines which can take their place at a workbench or a machine station alongside a human worker, occupying the same space and operating within a similar radius of action. Since robots of this kind are designed to do work that is intricate rather than particularly strenuous, and need dexterity rather than great reach or flexibility, their anatomy is light and they tend to employ a variety of specialist end effectors rather than general purpose grippers. In functional terms they are intended to be the equivalent of a human craftsman sitting at a workbench with a whole rack of power tools at his disposal and, physically, they rather closely resemble a drill stand or a dumb waiter.

At the other extreme are the robots which are, for one reason or another, abandoning anthropomorphic geometry or human scale. The range of physical dimensions has already widened to include both the very small and the very large. A Swiss firm has developed miniature robots that squat on a workbench and assemble watches like some race of ingenious mechanical gnomes, and a Scottish company builds giant robots which man-(or machine-)handle steel ingots weighing up to 20 tons.

But differentiations of size may be only the start. A robot with a circular radius of action about a fixed point, for example, is fundamentally ill-suited to work on a production line, where the ideal is a linear robot. It is possible to make the arm track a workpiece such as a car body as it moves along the line; and it is also possible to move the whole robot along a parallel line, though that involves shifting a substantial piece of engineering hardware back and forth. But a more elegant solution is to 'distribute' the robot along the line, perhaps giving it a single brain but a whole series of arms and hands, like the legs of a millipede.

It is also entirely possible that in the industrial setting the robot will, on the physical level, abandon all its anthropomorphic aspects. Once automation reaches the point at which it can dispense with human supervision altogether, factories need no longer be designed around human dimensions – if an industry is producing very small objects then it will make sense to have a very small factory where distances between machines are reduced to the minimum and the roof is lowered to a height of two or three feet. At the moment the industrial setting is still determined largely by human needs – workbenches, tools and access to machines are at waist level, overhead cables or machinery must be high enough to be over human heads and space must be provided for human workers to move around – but remove the humans and all the machines, including the robots, might just as well work at floor level and as close together as possible.

An even more radical suggestion calls for rethinking the whole concept of replacing the human worker with a mechanical surrogate which operates in analogous fashion. According to this argument, the anthropomorphic robot is a compromise forced upon us by the imperfections of contemporary automation and, when we reach the stage of complete automation and banish men from the workplace altogether, the need for machines which work like men will also disappear. The robot will not necessarily become extinct, but will be subsumed into a whole manufacturing system which could be viewed as a gigantic robot, a robot system in the same sense as a pilotless plane under autonomous control is a 'robot plane'. But within such a system there would be no need for anthropomorphic machines because the tasks which men, or robots, currently perform would be dealt with in quite different ways. An anthropomorphic robot is required to load a machine tool or stack a pallet, the argument goes, only because machine tools and pallets have been designed to be loaded or stacked by men. If the ground rules are changed, quite different methods will be found, rather as the screw replaced the paddle wheel when it was realized that a steam engine did not have to propel a ship by mimicking the human action of rowing, or the turbine replaced the reciprocating engine when it dawned on engineers that an engine which transformed energy into rotation directly would be simpler and more efficient than one which had to translate lateral movement into rotation by means of a crank.

One area in which robots are likely to miss out the anthropomorphic stage altogether is mobility. For most purposes wheels are a more effective and simpler means of locomotion than legs. Some industrial robots are already mobile to the extent that they move back and forth on rails or gantries in order to service a line of machine tools, work on a number of parts in turn or track a moving production line, and limited movement of this strictly predetermined kind is relatively easy to add to a robot's existing repertoire of skills. 'Free' movement about the factory floor or the warehouse, or, for that matter, the home, will almost certainly have to await the arrival of relatively sophisticated vision systems – a robot weighing a ton or more moving blindly around the workplace would represent an industrial hazard of the worst kind.

In the meantime, the control problems that mobile robots will encounter when they are free to manoeuvre around crowded spaces are the subject of much research. But the first result of this work is likely to be the 'robotization' of light trucks and fork lifts used around the workplace rather than the mobilization of existing robots. Some fairly elementary 'robots' of this kind are already used to sweep supermarket aisles at night, following magnetic tracks laid in the floors. Such machines will be robots in the sense that the human driver has been replaced by an electronic control system: there is obviously little point in designing an anthropomorphic machine which will sit in the driver's seat and push the brake or turn the steering wheel.

The one field in which there may be some demand for robots with legs is that of space exploration. There may well exist terrains which might be impassable for a wheeled or tracked robot but surmountable by one with legs. But the two-legged, human model looks distinctly unpromising. Four legs, or even six on the insect model, are a much more feasible plan since they provide far greater stability, a better grip on difficult surfaces, and demand a far less complex sense of balance in the control mechanism. Interestingly, Russian scientists who were involved in designing a six-legged robot for their space program found, on comparing notes with biologist colleagues, that they had, in effect, reinvented the insect. In order to achieve maximum stability and manoeuvrability, the robot had to move its legs in exactly the same pattern as an ant or a beetle, ensuring that five out of its six feet were always on the ground at any one moment. A more frivolous parody of the natural world invented by Marc Raibert of Carnegie-Mellon University is the robot kangaroo, which hops from place to place on a single leg in order to help investigate the control problems posed by its inherent instability when at rest.

But the engineering of a robot bestiary, indeed robot vehicles of most kinds, is outside the scope of this book. Robot planes, cars or submarines will employ and develop the hardware used in existing planes, cars and submarines, while drawing, no doubt, on the anthropomorphic robot when there is a need to add manipulator arms and end effectors. The aspect of such vehicles which is of concern to the roboticist is not the physical shape or means of locomotion employed, but the control system, the brain which replaces the human pilot, driver or helmsman. If human control is dispensed with, then the mechanical control which takes over will need to be equipped with the same set of senses (or substitutes for them), the same skills, experience and ability to make judgements and decisions as a human controller. None of these things is remotely possible without a 'brain' of one kind or another, and it is, therefore, to the computer, the one available brain for robots of all kinds, which we now turn our attention.

4. A Calculating Mind

The kingdom of numbers is all boundaries
Which may be beautiful and must be true;
To ask if it is big or small proclaims one
The sort of lover who should stick to faces.

W H Auden, 'Numbers and Faces'

The construction of robot brains poses problems that differ in kind as well as in degree from those connected with robot bodies. The engineers of artificial muscle and bone employ well-established mechanisms and techniques to reproduce the workings of a model, the human arm, of which they have a full description and clear understanding. The builders of brains, in contrast, rely upon a category of machinery, that dealing solely with information, which is comparatively novel and they have, in practice, been limited to the use of a machine of a single kind – the digital computer. Moreover, they have no blueprint of the model they seek to imitate, the human brain, and, therefore, no way of knowing whether the means they employ are adequate or even appropriate to their goal.

No one would suggest that a robot arm is an exact imitation of a human limb, but knowledge of engineering, on the one hand, and of anatomy on the other, enables us to define very precisely the ways in which they are comparable and the ways in which they are not. This, in turn, allows the roboticist to work confidently when he seeks a mechanical equivalent for a particular human skill or attribute. Indeed, it frequently makes it possible for the robot to improve upon the natural model, for technology has been able to better nature in many respects when dealing with physical movement. But even when the balance of advantage tips in favour of the mechanical, the relationship between the two systems is clear. It is evident, for instance, that a single degree of freedom in a robot elbow parallels the human model, a wrist joint with 360 degree rotation improves upon it, and an arm which extends and contracts telescopically achieves some of the same results by quite different means.

Physiology, which provides robotics with a clear and unambiguous picture of the way in which the limbs function, can offer little more than conjecture when it comes to the brain. The study of dead tissue tells us a strictly limited amount about how or why the living brain works, and there are obvious practical and ethical limitations to research on living specimens. There is also a philosophical limitation to the brain's introspective study of its own workings, in that there is always a suspicion that conclusions will be coloured by some innate disposition in the brain itself or, to put it another way, that the brain's perception of its own workings may be neither accurate nor complete.

Knowledge of the human brain is currently confined to two fragile bridgeheads bordering on a central gulf of ignorance. On the one side, neurophysiology can give a superficial picture of the structure of the brain and offer some educated guesses about its operations; on the other, psychology provides a variety of hypotheses about its output. How, exactly, the one gives rise to the other remains mysterious. For robotics, the problem can be compared to that of trying to fathom the principles of the internal combustion engine by examining the exterior of one, snipping a lead or severing a pipe here and there to see what happens, and observing that a block of metal, which one lacks the tools to dismantle, somehow makes the wheels go round. Trying to build another engine based on this level of information would clearly be a hazardous undertaking.

In the absence of sufficient knowledge about the natural brain on which to base the design of artificial ones, robotics has been forced to rely upon the only available machines which can plausibly be said to fulfil at least some of the criteria which define a brain. Whether the computer is capable of meeting all the needs of robotics, or, perhaps more importantly, providing a model on the basis of which comparisons between natural and artificial systems can be made, is very much an open question. But it is indisputable that it was the invention of the computer which made robotics possible, and that it has provided a more than adequate brain for the kinds of robots that are in use today. The best starting point for a discussion of robot brains is thus to attempt to answer the question: what kind of machine is a computer (or, to be more precise, an electronic, stored program, digital computer)?

The attribute which distinguishes the computer from most of its mechanical forbears, and the one which it most obviously shares with the brain, is the ability to operate on information as an abstract entity detached from any physical manifestation. This being not altogether an easy concept to pin down, one might borrow the usage that lawyers employ when they talk of 'incorporeal property' as opposed to physical property. While it is perfectly possible, and often convenient, to define a piece of physical property, such as a plot of land, in terms of information (marks on a map for instance), that information exists only as a description of a physical entity; on the other hand, incorporeal property, the copyright of a literary work, say, or a musical composition, is in itself a piece of property regardless of whether it is given any physical form as a book or a recording.

Mathematical or logical information clearly has this property of being abstract or 'incorporeal'; the number 17 exists independently of its attachment to physical objects, like 17 floors, or measurements, such as 17 hours, and the same is true of logical relationships like 'and', 'equivalent to', 'if', and 'or'.

A computer (or a brain) has the ability to assimilate, store (or 'remember') and process (or 'think about') information in the 'incorporeal' sense, and this separates it from virtually all other machines which operate in the medium of information. There are many such machines which deal with information in a less abstract sense. Clocks, weighing machines or thermometers measure information, telephones or radios transmit it, tape recorders commit it to memory. Yet these machines usually work in a way that is specific to information of one particular kind – movement, sound waves, etc – and the information they deal with consists of measurements of one sort or another. If, for example, a tape recorder 'remembers' a piece of music, what it is 'remembering' is a description of the sound that music makes, and the music on the tape is not music in the sense of, let us say, the score which the composer wrote. On the tape, the C sharp in the second bar is a particular C sharp played on a particular instrument and exists only for the time the note was held; the C sharp in the score exists as information, which may be translated into sound by any instrument, as long as a copy of the score remains in being. Unlike the tape recorder, the computer operates with information in the second category, one which clearly has more of the quality of ideas than of things.

Yet, if a machine is to assimilate, store or process an 'idea', some way must be found to give it a physical existence in terms of 'things'. The obvious way to achieve this is to use the *state* of the machine or its components to represent the information. All measuring machines do this: a clock represents information about time by the state of its hands; a thermometer, information about temperature by the state of the mercury it contains. But, clearly, the states used by these machines are very specific to the kind of information they are designed to deal with – a clock cannot deal with information about temperature or a thermometer with information about time. Moreover, most such machines have an infinite range of states which only yield up the information they represent if some process of measurement is applied from outside or if there is some scale or calibration against which the state of the machine can be measured.

This fact can be illustrated by looking at a simple example. An old-fashioned ship's telegraph is a device which uses the state of a handle or lever to transmit and store information about the speed at which the captain on the bridge wants the engines to work. In theory, the state of the system could be used to represent any gradation between 'full ahead' and 'full astern'; in practice it is necessary to divide the scale up into arbitrary divisions denoting 'half ahead', etc. This principle of having a limited number of 'discrete states' is common to most machines which are designed to deal with information. Though all possible physical states of a machine may form a continuum which is infinitely divisible, it will be necessary to limit the number of significant states, the ones which represent or convey information, to a finite number and separate them from each other so that each state represents a definite, or digital, quantity of information. The need to separate abstract information into discrete chunks inevitably leads to the idea of dividing it mathematically, since mathematics deals in numbers, which abstract properties common to many different kinds of information. In principle, mathematics is concerned with 'real numbers' of infinite precision, like *pi*, but in practice even mathematicians will often find it convenient to treat them as digital quantities.

But, though we are accustomed to thinking of the physical world in numerical, mathematical terms, since it would otherwise be impossible to attach any meaning to ideas like the distance from London to New York or the time taken to travel to work, we instinctively resist the idea of treating abstract thought or feeling in the same fashion. It would be repugnant, for example, to measure political convictions, emotional attachments or even a preference for roast beef over grilled fish in mathematical terms; and we usually blur the issue, though not perhaps the principle, by measuring such things in vaguer terms – 'preferring' this candidate to that, loving someone with 'all our heart', liking beef 'better' than fish. Builders of brains, however, are not alone in finding mathematics to be the only logical starting point for dealing with abstract information. Philosophers from Plato onwards, seeking some bedrock on which to erect structures of thought, have been attracted by the certainties of mathematics which, starting from facts which seem self-evident and are confirmed by the senses, can reach conclusions which are far from obvious and highly abstract. It is hard to think of another mode of thought which can, for example, start with a 'fact', such as $1+1=2$, which is easily demonstrated in the physical world and arrive at an idea, the square root of 2, which can only exist as a concept.

The idea of a mathematical machine or calculator did not, of course, originate with the computer. The first mechanical calculators were built in the seventeenth century by, interestingly enough, philosophers. Pascal designed the first adding machine and Leibnitz the first multiplier, and the principles they employed remained in use until the advent of the electronic calculator in the early 1970s. The central idea is that of using a series of wheels on a common axis to represent, by their state of rotation relative to some fixed point, a number in the traditional decimal notation. A familiar example of the principle is the mileometer on a car which, normally, has six wheels, each divided into segments numbered from one to ten and interlinked so that the 'tens' wheel, for example, advances by one segment for each complete rotation of the 'units' wheel. Such a device has 1,000,000 possible states, representing distances from zero miles to 99,999.9 miles in digital units of one-tenth of a mile.

Though a mileometer counts rather than calculates, it is not difficult to envisage a mechanism which would enable it to add and subtract more than one unit in a single operation and, with some ingenuity, it is possible to add another which enables it to multiply and divide; though it is worth noting that many mechanical calculators accomplish these latter two functions by treating them as repeated processes of addition or subtraction.

The wheel, however, is not a device which is inherently digital; it requires engineering skill to confine it to a set number of states, for it more 'naturally' revolves smoothly than in a series of clicks. Also, its state will only have any value as information if it is measured against some fixed point – take a mileometer out of its casing and the state of the wheels could be interpreted as any one of ten different numbers. A switch, on the other hand, is digital in a fundamental sense; its function requires that it have two, and only two, states – on or off, open or closed. It also has the advantages that, if it is treated as a device for allowing or preventing a flow, it needs no external datum point against which to measure its state, and series of switches can be interlinked or combined more easily and flexibly than combinations of wheels by transmitting flows from one switch to another.

For all its digital qualities, the switch has obvious drawbacks when applied to the decimal system. It will require ten switches rather than one wheel to store, in terms of their state, a number between 0 and 9. It is quite feasible to build machines which overcome this difficulty; indeed the machine generally considered to be the immediate predecessor of the computer, the ENIAC, used switches in this way to represent decimal numbers. But the decimal system's base of ten is incidental rather than fundamental to mathematics. It is perfectly possible to count and calculate using any number at all as a base, and the switch obviously lends itself to the use of a base of 2, the binary system, which requires only two digits, 0 and 1 – on and off. This has advantages when dealing with information, since on and off can represent yes or no, true or false as readily as 1 or 0.

To non-mathematicians, the strings of 1s and 0s which make up binary numbers often appear cumbersome and complicated by comparison with the familiar decimal notation. It looks far easier to write or read decimal 15 than binary 1111 (for those unfamiliar with binary numbers it may be easiest to think of 1111 as, reading from the right, one 1, one 2, one 4, and one 8 in exactly the same way as one considers 15 as five 1s and one 10). But if one looks at the implications of encoding numbers in terms of the state of a machine, it becomes clear that the reverse is the case.

Take, for example, the number 98, or 1100010 (no 1s, one 2, no 4s, 8s or 16s, one 32 and one 64). To represent 98 in a traditional calculator requires two wheels, each with ten permitted positions, or a machine which has to have a repertoire of 20 possible movements from state to state in order to achieve a total of one hundred states. To represent the same number in binary digits takes seven switches, each with two positions, or a machine with a repertoire of only fourteen movements which can be combined to produce 128, or 1111111, possible states. The switches will, of course, require more interlinking connections than the wheels, but the way in which they are achieved is fundamentally simpler. A simple formula, which it is useful to remember, states that if 2 is raised to the power of the number of digits in a binary number the result will be the largest number which can be represented by that number of digits. Thus a two-digit number can represent any number between 1 and 4 (2^2), a three-digit number anything up to 8 (2^3) and so on. If we think of different numbers as conveying different messages, then we can say, for example, that a combination of eight binary digits can be used to convey 256 (2^8) different messages.

A further advantage of the binary system is evident when it comes to multiplication. In decimal arithmetic, the multiplication table is a grid of one hundred squares showing the products of every possible multiplication of the digits from $0 \times 0 = 0$ to $9 \times 9 = 81$; in binary arithmetic it is a grid of only four squares, thus:

×	0	1
0	0	0
1	0	1

The binary multiplication grid

The benefit of this simplicity becomes clear when one compares the process of 'long multiplication' in the two systems. Say that the problem is that of multiplying 23 by 36. In decimal, the first stage is to multiply 23 by 6, remembering to carry a ten from the units column to the tens and a hundred from the tens to the hundreds, to get 138; then 23 is multiplied by 3 after adding a zero to allow for the additional order of magnitude, to get 690, and finally the two partial products are added together to get the result, 828. In binary, multiplying the same numbers, 10111 and 100100, requires only a simple addition. For, ignoring those stages where the multiplier is a zero, the products of a multiplication would be 1011100000 and 1011100, or the multiplicand multiplied by 1 with appropriate zeros added. Performing the addition:

$$\begin{array}{r} 1011100000 \\ 1011100+ \\ \hline 1100111100 \end{array}$$

demonstrates another factor: carrying is reduced to a simple rule of thumb. $0+0=0$, $1+0=1$ and $1+1$ gives 0 and a carry to the next column. This can be represented in another grid:

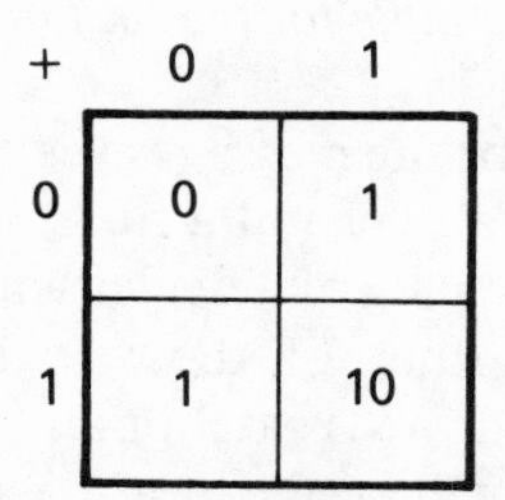

The binary addition grid

Looked at in this way, binary arithmetic can be seen to be a logical as well as a mathematical process, for if, going back to the multiplication grid, 0 is replaced with 'false' and 1 with 'true' it becomes what logicians call a 'truth table' which shows when, given two propositions, A and B, both are true:

AND	False	True
False	False	False
True	False	True

Truth table for the AND *function*

In effect, the function AND has replaced the function *multiply*. In the same way, the grid for addition can be thought of as utilizing the function OR in the sense that it tests whether proposition A *or* proposition B is true:

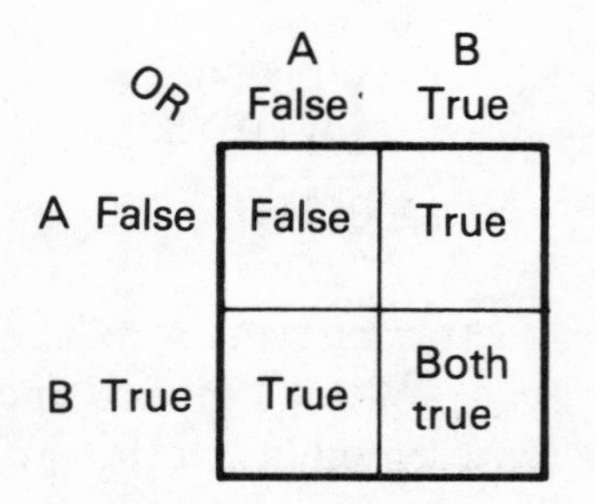

Truth table for the OR *function*

If the digits 0 and 1, or the messages yes or no, take the form of a flow of current, or the absence of a flow, which operates a switch, and if the switches form part of a device which generates new flows according to one or other of these logical rules, one has, in computing language, an 'AND gate' or an 'OR gate' and combining these in various ways will result in a machine which 'does' both binary arithmetic and logic. It will in fact incorporate what are known as Boolean functions, after the British mathematician George Boole, who in the nineteenth century produced an algebra based on grids like this which he proposed, rather optimistically, as a model of the 'laws of thought'.

The first electrical circuits which worked on Boolean principles were designed in 1938 by an American, Claude Shannon. Today, 'logic circuits', made up of AND gates, OR gates, NOT gates (which transpose a 1 into a 0 and vice versa) and further elaborations like 'exclusive or' (EXOR) gates, still form the heart of every computer, the parts that enable it to compute. The way these gates work can be shown by a very simple diagram (opposite).

We have seen, thus far, that a computer is a machine which stores the simplest kind of digital information (units and zeros) in the form of the states of combinations of switches and which can manipulate or combine two bits of information according to rules which, though they can be used to perform mathematical tasks, are in essence an expression of logical relationships. In fact, the most sophisticated mathematical rules that a computer employs, at the fundamental level of its logical circuits, are $0+0=0$, $1+0=1$, $1+1=10$, $0 \times 0=0$, $0 \times 1=0$ and $1 \times 1=1$. Such elementary

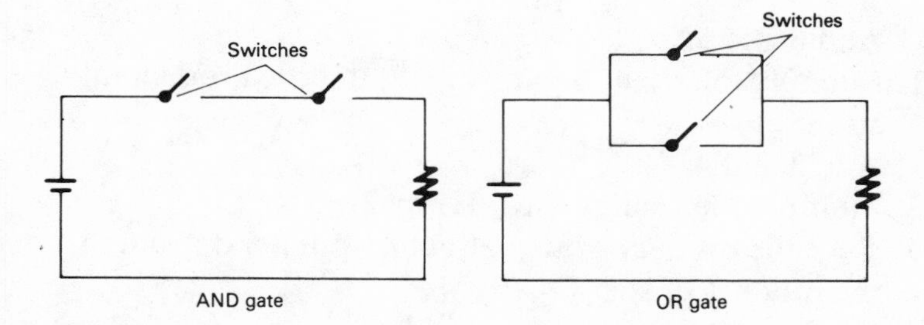

Schematic representations of two kinds of gate. If it is imagined that the switches are operated by incoming digits (a 0 causing the switch to open and a 1 causing it to close) then it can be seen that, in the case of the AND gate, a current will flow around the circuit from the battery on the left, creating an output (or a 1) on the right, only if both digits are a 1. However, in the case of the OR gate, a 1 will be output if either of the inputs is a 1

forms of 'memory' and 'thought' seem, at first glance, irreconcilable with the complexity of the brain, and taken in isolation this is clearly so.

But a machine which simply adds two numbers or multiplies them together, whether in decimal or binary arithmetic, and whether by rotating wheels or operating switches, may be convenient, but it can scarcely be said to be doing mathematics in anything but the most trivial sense. Any worthwhile mathematical problem is likely to involve a series of computations; even a simple equation like $x = 3(6 + 5) - 2(3 + 4)$ requires two processes of addition, two multiplications and a subtraction, all to be performed in the correct sequence, with intermediate results set aside until they are required. If a human mathematician performs a calculation of this kind with the aid of a simple calculator then it could be said that it was he who was doing the mathematics, not the machine, and he would require, if the numbers were of any size, a pencil and paper to jot down the results of each separate stage. A machine that could undertake a whole sequence of calculations would seem to come much nearer to 'doing' mathematics in the same sense as a mathematician does them. But it would require three things that are lacking in a calculator: instructions for each step, a system of control to ensure that they are taken in the right sequence with the right numbers and a memory to store intermediate results. Solving the equation above would, for example, require instructions as follows:

I Add 6 and 5.
II Multiply the result of step I by 3, put the product into store.
III Add 3 and 4.
IV Multiply the result of step III by 2.
V Take the product of step II out of store and deduct from it the product of step IV.

If more than one intermediate result is involved, each will have to have an address in the memory store, and the machine's controls will have to keep track of which number is at which address.

But although a machine which can carry out a process like this would be a major advance, in that many kinds of problems can be dealt with by simply following a set of instructions from A to Z, there are other problems in which the nature of the next step can only be determined when the previous one has been completed. Even simple, apparently mechanical, processes like book-keeping will require such 'decisions' to be made. For example, a clerk, or a machine, might be asked to add up two columns of figures to arrive at two totals, A and B, and then, having ascertained the balance between them, C, to add C to one or other of two further totals, the choice being determined by whether A was greater than B or vice versa. The instruction for a calculation like this will have to include the word 'if'. At one level, the 'if' function is a part of the Boolean system; the grids, or the circuits which form mechanical substitutes for them, are essentially ways of resolving 'if' questions by rules such as 'If 0 and 1 then 1 but if 0 and 0 then 0.' In fact, when the Boolean system is realized in the form of electrical circuits, this capacity for dealing with what we shall call 'if . . . then . . . ' problems is even more entrenched. Shannon showed in 1938 that any arrangement of his logic circuits which makes provision for 'feeding back' the result of one computation and treating it as a component of the next will result in a device where the 'if' function is a fundamental property of the system.

We can illustrate this point by looking at the arrangement in the diagram opposite. Here an EXOR gate (the function of which can be summed up as: 0 and 0 = 0, 1 and 0 = 1, 1 and 1 = 0) has been wired up so that its output returns, after what is called a 'unit delay' (we will come back to this in a moment) to provide one of its inputs.

Suppose that the system starts with inputs of 1 and 0; its output will be a 1. This output, transformed into an input to the next cycle, has the effect of determining the gate's function when it receives another, single digit, input. *If* the new digit is a 1, the new output will be a 0 but *if* the new digit is a 0, the output will be a 1. The point is that the gate itself now has a 'state' which gives it the property of outputting a 0 when the input is a 1 and vice versa.

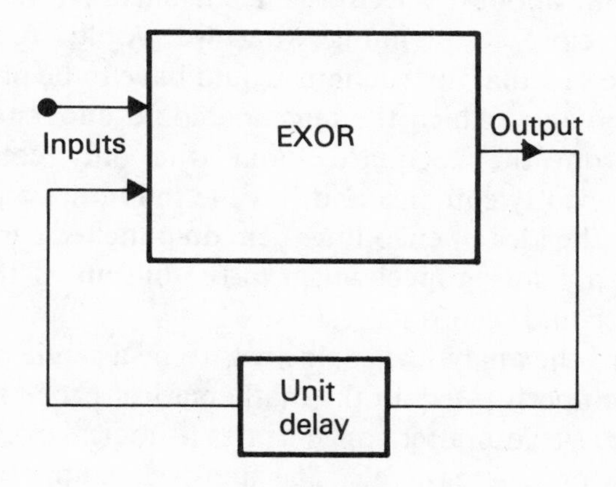

Gate with feedback

The importance of the unit delay is that if a machine is to have discrete states its workings must go in stages, like the frames of a movie film passing through a projector. If the feedback flowed as a constant current, the device would be much harder to control since it would be tricky to pin down its state at any particular moment. Some 'asynchronous' systems do exist for specialized purposes (in communications and plant control, for example), but a *digital* computer must be 'synchronous'. In practice, the discreteness of the states of all its components, whether switches in the memory or logic circuits like this one, is ensured by the 'clock pulse', a current of electricity which flows to all parts of the machine simultaneously, perhaps hundreds of thousands of times a second. While the pulse is present, the machine as a whole is in one state, just as the frame of a film must be stationary in the projector gate for the instant during which the shutter allows light to pass through it.

The concept of a machine which could follow a whole sequence of instructions, including those which involved 'if' functions, first arose in the mind of a British mathematician of the early nineteenth century, Charles Babbage. Babbage had, perforce, to base his design on the technology of wheels and gears employed in mechanical calculators (the idea of using switches or binary arithmetic was still a century away) and his plan proved impractical if only because contemporary engineering was incapable of producing components with the necessary degree of precision. But though his 'analytical engine' never came into existence, this should not disguise the fact that, in thinking about the needs of such a machine, Babbage came astonishingly close to planning what we would, today, call a computer. He saw that instructions would have to be provided in a code, or 'language', which the machine could understand and he borrowed, from the Jacquard loom (the only contemporary machine sufficiently complex and flexible to require a program of instructions), the idea of encoding them on punched cards. He also incorporated a printing mechanism in the output of the machine with which it could record its results.

The heart of the analytical engine was to be a single mechanism, the mill, which performed all the mathematical processes in much the same way as a computer concentrates its logical operations in a central processor. There was also an extensive memory, in the shape of rank upon rank of gearwheels, for storing intermediate results and a control system to ensure that the mill moved the right numbers in and out of store at the right time and performed the right operations on them. In a final, surreal touch, Babbage proposed to power the whole contraption with a steam engine. As was probably inevitable, he died with his grand design unrealized, awaiting a technology which made it practicable. The next time the idea of an all-purpose machine surfaced it was approached from a very different point of view, that of pure mathematical speculation untainted with concepts of utility or feasibility.

The Turing machine was, and remains, a completely abstract idea. Its inventor, the British mathematician Alan Turing (whom we mentioned earlier as the inventor of the Turing test), was doing postgraduate work at Cambridge University in the 1930s when he became interested in the so-called *entscheidungsproblem* (decision problem) posed by the German mathematician David Hilbert earlier in the century. The question at issue was whether mathematics was 'decidable', whether there was some definite method which was capable of resolving the truth or otherwise of any mathematical proposition. To put it less formally, the question was whether there was such a thing in mathematics as an insoluble problem.

The first step was obviously to try and define a definite method which comprehended all possible forms of computation and calculus from simple arithmetic to the most arcane algebras. Turing came up with a machine. Not a machine that it would necessarily be practicable to construct, but a machine that could be described in terms that made it clear that everything it did would be entirely 'mechanical'. The machine Turing envisaged consisted of a 'head' moving backwards and forwards along a 'tape' which was divided up into squares. In order for the machine to be capable of every kind of calculation it is necessary to assume a tape of infinite length, though any given calculation would take only a finite number of squares. The head moves along the tape, one square at a time, to left or right, and each time it moves it 'scans' a square for any symbol it may contain. The head may leave a square unchanged, whether it contains a symbol or not, erase a symbol and leave a square blank or replace it with another symbol, or may write a symbol in a blank square. Though symbols of any kind can be used, any calculation will require only a defined range, that is to say the machine only uses the 'alphabet' it is 'given' when a problem is set out on the tape.

The action the head takes at each juncture is determined by a combination of two factors: what it finds in the square that it is scanning and what Turing called the head's 'configuration'. The configuration is essentially the same thing as a 'state' but in the context of a Turing machine it helps to think of the head having a state in the sense that a car gearbox is in a particular state when a certain gear is selected. For the state of the head determines the output, or action taken, in response to an input, or symbol scanned, rather as the gear determines how the car will move when the clutch is released.

Suppose that a Turing machine is required to add 2 and 4 together. The problem will be set out on the tape in the following form:

11 1111

with the head resting somewhere to the left of the first cluster of symbols. The instructions given to the machine will tell it to move to the right in State 1. In this state the head moves, scanning each square, until it encounters the first square containing a symbol. When it scans the first 1 its instructions tell it to change into State 2, in which it will continue to move to the right scanning each square until it encounters one which is blank; it will then write a symbol in the blank square, resulting in a tape like this:

1111111

Its instructions then tell it to change to State 3 in which it moves again to the right, scanning the squares containing the second cluster of symbols but leaving them unchanged, until it again encounters a blank square. It leaves the second blank unchanged, but itself changes to State 4, in which it moves one square to the left, scans the symbol it contains and deletes it. It then comes to rest, its task completed, leaving the answer like this: 111111

If this seems so elementary as to be pointless, it may be worth looking at another problem, which shows that the Turing machine head has states or configurations which deal with the 'if' function. Imagine that it is asked to decide whether some sequence of letters such as ABBCDDCBBA is palindromic (that is, that it reads the same from left or right). Starting from a position to the left of the cluster of symbols, the head is instructed to move to the right until it encounters a symbol, which it then deletes and replaces with, say, an X and enters a state in which it continues to move right until it finds the blank square marking the end of the sequence. It then changes state again and moves left to scan the last symbol; *if* this is not the same as the one it has deleted at the other end of the cluster, it stops, having shown that the letters do not form a palindrome, but *if* it finds, as in the example, that the symbol is the same it deletes it and replaces it with another X. The head then continues to shuttle back and forth replacing symbols with Xs, until it finds a symbol that does not fit the pattern or until it has replaced all symbols with X, leaving the tape marked XXXXXXXXXX.

Another way of thinking of the configuration, and one employed by Turing himself, is as a summary of progress made to date on the calculation which is in hand. For it can be seen that the instructions do not tell the machine what to do in each square: they tell it what state it should be in and which way it should move, and it is the combination of configuration and symbol scanned which determines what the head does. A particular configuration, therefore, can be thought of as a rule which tells the head what to do with each blank or kind of symbol it next encounters, a rule which effectively encapsulates the state in the calculation which has been reached.

Turing was able to show that a machine of this kind could, in principle, tackle any possible calculation for which a clear method could be provided in the shape of a table of instructions. The technical term for such methods, or the instructions which describe them, is algorithms, a word we shall meet throughout the remainder of this book.

Another way of describing a machine which can, in principle, tackle any and every problem is to say that it is a 'universal' machine. Actually to solve a problem of any length or complexity with a Turing machine would be impractical, but the perception that every question, however large and elaborate, could be dealt with in this way, one step at a time using only elementary methods, was a crucial advance, and the invention, if only in theory, of a universal machine can be seen, in retrospect, as an epoch-making event.

To resolve the question that concerned him, Turing took things a stage further. Given that the machine could solve any possible problem that could be defined, it would in theory be feasible, he argued, to list all possible problems in a numbered sequence, and to attach to each entry in the list the table of instructions which told the machine how to tackle that particular problem. Further, since the number attached to each table of instructions could be thought of as a code denoting that set of instructions, translating the code into a table of instructions was simply another kind of calculation which could, in principle, be entrusted to the machine. It was, thus, possible to conceive of a machine which, given a list of numbers representing all possible calculations, would plough its way through, decoding each number into a set of instructions which it would write on its tape, read and act upon until it had produced a tape cataloguing all possible problems and their solutions – all, as Turing put it, 'computable numbers'. This illustrates another important point. The 'universality' of the Turing machine is a consequence of the fact that it is only when it has a set of instructions that it 'becomes' a machine of a certain kind (the kind required to carry out those instructions); if another set of instructions is substituted, a different kind of machine takes its place.

A further point is that, by taking the step of encoding the instructions to his machine in a form of calculations it itself could perform,

Turing was proposing that it store and reconstruct its own program, a concept that runs counter to the common sense notion that data and instructions are distinct and separate categories that must be provided to a calculating machine in different forms. The technique of using numbers to define mathematical processes or, in Turing's case, instructions to carry out processes, had become commonplace in pure mathematics – Gödel's famous theorem proving the incompleteness of mathematics rests upon it; but the idea of applying it, even in theory, to calculating machines was totally novel and, at first sight, nonsensical. However, when applied in the computer, it proved to be the key which opened the door to a mechanical brain, of one kind at least.

The resolution of the *entscheidungsproblem* was achieved by means of an established mathematical 'trick'. The list of all possible tables of instructions would, clearly, be infinitely long, and if the number attached to each was also infinitely long, even if only by dint of adding zeros, then it can be shown that any such list of infinitely long numbers can give rise to another number, the Cantor diagonal number, which is not in the list. This number would define an instruction table for a problem that was not in the list of all possible problems and was therefore, as Turing put it, 'uncomputable'. Turing's paper, 'On Computable Numbers . . . ', which described his machine, is reckoned to be the most significant landmark in what might be called the prehistory of the computer, for reasons which will be clear if its method of working step by step on one symbol at a time is compared with the very similar methods used by the computer. He showed that a machine which operated in this apparently simple-minded way could, in principle, be universal; and thus demonstrated the potential of a computer before such a thing existed.

Turing's conception is, rather obviously, not a blueprint for a practical working machine in the sense that Babbage's was, and the computer, the first working universal machine, came into existence to meet urgent worldly needs rather than to satisfy disinterested curiosity about abstract mathematical problems. Many aspects of the Second World War – the need to calculate artillery range tables, codebreaking, and, ultimately, the building of the first atomic bombs – involved calculation on a scale which threatened to outstrip human mathematicians, not because the problems were necessarily very complex, but because they were very long. Producing range tables, for example, involved calculating the effects of variables like elevation, propellent charges and wind velocity separately for each tiny portion of a shell's trajectory.

Before the outbreak of the war it had become clear that the switch offered advantages over the wheel in the design of digital 'state' machines, and the most promising kind of switch seemed to be the electromagnetic relay (described in Chapter Two) which was a standard component of telephone and telegraphic systems. It operated at far higher speeds, ten or more times a second, than purely mechanical switches, and messages travelled from switch to switch instantaneously in the form of an electric current. In the late 1930s a German engineer, Konrad Zuse, had built a calculator which used electromagnetic switches to perform binary arithmetic and in 1937 an American, Howard Aiken, launched a far more ambitious project for a giant machine of this kind, the 'Harvard Mark I' as it became known. This machine used relays on an heroic scale, filling a vast room with cabinets of them clicking away like some manic knitting machine. It was not, and was never intended to be, universal; its main task of calculating range tables for the US Navy required it to perform repetitive calculations at high speed rather to undertake a wide variety of jobs, and by the time it came into service in 1944 it was already being overtaken by events.

The war not only demanded that unprecedented feats of calculation be performed with extreme urgency, it also stimulated interest in a technology that was to prove crucial to the process. Electronic devices were not a novelty in the 1940s, but until the British began the program of research which led to radar, their use had in practice been largely limited to radio. However, the valves, or vacuum tubes, which, in small numbers, were used as amplifiers in radio sets could also be thought of, when used in combinations, as switches, operating almost instantaneously in cycles which could reach a rate of a million per second. The difficulty was that valves were extremely expensive, consumed a great deal of current, generated a lot of heat, and, conventional wisdom had it, were so delicate and unreliable that any machine depending on large numbers of them operating in concert would be doomed to failure.

The success of radar, and the growth of a substantial body of practical experience which resulted from its development, did, however, produce an atmosphere in which electronics became, so to speak, a fashionable technology; if a problem existed it was tempting to see if there might not be an electronic solution to it. The war was, moreover, a period when conventional wisdoms were questioned daily and the solution of scientific problems became, literally, a matter of life and death.

So, some time before Aiken's machine was complete, other researchers at the Moore School in Philadelphia had taken the plunge and embarked on the construction of an entirely electronic machine. The ENIAC (Electronic Numeral Integrator And Calculator) can be considered the immediate ancestor, if not the prototype, of the computer despite the fact that it was somewhat behind the times and worked in decimal rather than binary arithmetic, using a group of ten valves to represent each digit.

In principle, ENIAC approximated the ideal of a universal machine in the same sense as Babbage's analytical engine: its valves provided storage sufficient for it to undertake very ambitious calculations and, of course, it reduced the time scale by a factor of tens of thousands, since its electronic state could alter hundreds of thousands of times a second. Yet the very speed at which it operated raised difficulties that detracted from its universality. To feed instructions to such a machine on punched cards, or by any other means that had to be read by intermediate, mechanical operations, would have been colossally wasteful, for no mechanical input system could keep pace with the machine's internal workings. The solution the ENIAC's designers came up with was a plugboard, similar to the pinboard which controlled the movements of the early robot manipulators. The instructions for dealing with each kind of problem had to be worked out in terms of interconnections between the different parts of the machine, which were then made by laboriously plugging in leads on the board. Once the plugboard was set up, ENIAC could tackle any problem which fitted that pattern with a speed which far outstripped its rivals; in practice, this was valuable since the calculations involved in, say, the compilation of range tables are repetitive to the point of monotony, but it placed ENIAC in a curious limbo, somewhere between the special purpose calculator and the universal computer; it was, in a way, not a single machine but a kit from which a whole variety of machines could be assembled. It was a kind of Turing machine in which the process of 'redesigning' the machine for a new task involved much arduous replugging rather than simply the provision of new instruction tables.

In Britain, the impetus for the development of new forms of calculating machinery came largely from the secret code-breaking establishment at Bletchley Park. Here the requirement was not for general purpose, mathematical machines, but for very special purpose ones; yet the breakthrough came in a similar fashion, when the decision was taken to ignore accepted views about the unreliability of the valves and try to see if large scale electronic machines could work. The results, the 'Colossus' machines, were special purpose calculators with no aspirations to universality, but, like ENIAC, they convinced those who saw them, a small but privileged group which included Alan Turing, that electronics offered a means of revolutionizing mechanical calculation.

If logical operations could be performed at rates of up to one million per second, then even the elementary, painstaking step-by-step processes of something like a Turing machine could take place at a rate which gave them practical value. Complicated problems might not have to be tackled by complicated and incomprehensible processes, but could yield to series of extremely simple ones carried out at astonishing speeds. This line of thought, which naturally leads to the hypothesis that thinking itself might consist of an accumulation of very elementary steps, was encouraged by the evidence that suggested that the brain itself was an electronic, digital, discrete state mechanism. In 1943 two American researchers, McCulloch and Pitts, showed that the neurons, or nerve cells, which are the basic components of the brain, are functionally very similar to the valve circuits which provided the switches in machines like ENIAC and the Colossi. Each neuron can 'fire' or transmit an electrical current which is received by other neurons via 'terminals' called synapses, and the currents take the form of discrete, digital pulses. The web of intercommunications linking the neurons is immensely complex, and looks nothing like the ordered bundles of cables which made up the circuits of the calculating machines, but it seemed clear that in some fundamental way the two systems were comparable and that if enough electronic switches were assembled and interconnected in the same way as neurons in a brain, they ought to behave like a brain. There are something of the order of ten thousand million neurons in the brain, and the number of possible states will therefore be incomprehensibly large, but the number of meaningfully distinct states will be finite.

The most striking difference between the brain and a universal machine which employed technology of the ENIAC kind is the fact that the machine was totally reliant on instructions provided from outside; it might be capable of solving any 'computable' problem, but without a human brain to tell it how to do it, it was helpless. Such a machine is able to 'do' mathematics, but the brain 'knows' mathematics, or, to put it another way, it contains the instructions for doing mathematics. In so far as those instructions consist of knowledge, like the right method for long division or the process for finding a square root, which have been acquired through training and education, the brain could be said to be storing programs of instructions which had been fed into it.

The idea that the instructions for a calculating machine did not have to take the form of punched cards or the arrangement of a plugboard, but could be fed into it and stored in terms of its internal states, probably occurred to several people at more or less the same time. It would have come naturally to Alan Turing who, after five years as a lynchpin of the code-breaking effort at Bletchley Park, turned his attention in 1945 to realizing his dream of a universal machine, which had seemed so impossibly impractical and remote only eight years previously. For the resolution of the *entscheidungs-problem* turned upon the idea that the Turing machine would write out its own instructions on the same tape as that which contained its data and its calculations. But the team led by J P Eckert and John Mauchly at the Moore School in Philadelphia which had produced the ENIAC had also been thinking about the next step, and they had the advantage of the services, as a consultant, of one of the greatest mathematical brains of the time, John von Neumann. In the event, the idea of a stored program calculator was first articulated publicly in the proposal which von Neumann drafted for the Moore School describing its next project, the EDVAC, and it is von Neumann who is generally credited with the crucial perception. To this extent, all subsequent computers could be called 'von Neumann machines', for the stored program was the development which transformed the calculator into the computer.

At first glance, storing instructions internally rather than taking them in from outside seems an advance in technique rather than in principle: the instructions will, after all, be the same, though it is undeniably more convenient if they are accessible in instantaneous, electronic form. It is tempting to draw analogies with our own experience, and feel that, while it may be very helpful for a cook to remember the recipe for sponge cake rather than having to look it up in a cookbook, it is scarcely fundamental. But to see the issue in these terms is to underestimate the extent to which our own brains are preprogrammed. A computer, faced with, say, two simultaneous equations, needs not only the algorithm for that kind of problem, but it also needs instructions for all the other basic processes involved in it – adding, subtracting, etc. Without instructions from outside or a program inside, it has literally no clue as to what it is supposed to do and, if the instructions are to come from outside, they will have to specify each step down to the last detail.

A calculator without a program is not, therefore, like a cook who cannot remember a recipe; it is like a cook who does not know how to measure the ingredients, how to beat an egg or how to switch on the oven. An unprogrammed machine requires each set of instructions to start from first principles and to be repeated each time they are needed.

The ability to store programs internally revolutionizes the nature of the machine, for it can then be equipped with a whole repertoire of programs to which it can refer, in any order, as and when it needs them. Moreover, large programs can be broken down into series of smaller ones, or sub-routines, which are 'nested' within the main program, so that the instructions for solving simultaneous equations do not have to be spelled out in painstaking detail, but can be built up out of existing programs covering the constituent processes.

Equipped with the facility for storing programs, the computer becomes a machine which can 'know' any process that can be reduced to a series of clearly defined steps. It is perfectly possible to view the brain as no more than a very elaborate machine of this kind. After all, the steps that are occurring in a computer's logic circuits are still the elementary, 'simple-minded' steps of the Turing machine or the Boolean grids; but supplied with instantaneously accessible instructions for performing them at a rate of hundreds of thousands per second in elaborate combinations, the machine as a whole seems far from simple-minded. Indeed, it is capable of feats that are quite outside the scope of the human brains which designed the circuits and drew up the programs.

Even the early primitive computers, with numbers of valve circuits and, therefore, a range of states, which were trivial compared to the millions of neurons in the brain and the untold billions of states they could, in combination, form (the number of states exceeds the number of particles in the universe), were capable of outpacing the brain; and it was logical to conjecture that this ability might not be confined to the realm of mathematics. The brain and the computer, after all, appeared to be the same sort of machine – a digital, discrete state, stored program machine. And though the computer had come into being as a specifically mathematical device, its mode of operation is, as we have seen, fundamentally logical in the sense that it does not incorporate the rules of any particular calculus, but manipulates information in its most basic form, single digital units, in order to simulate every possible calculus. It seemed possible, even probable, that the brain handled all information, not just mathematical information, in this way.

In the first flush of enthusiasm for the new-born computer, it seemed that making machines that 'thought' might thus come down to a matter of remodelling brain functions in terms of programs the computer could handle. If the brain and the computer operated on information in the same basic way, then enabling the computer to think about anything was a matter of finding the right algorithm, just as enabling a Turing machine to solve a particular problem was a matter of defining the table of instructions, for anything that a brain did must represent a 'computable' problem for a universal machine.

Many objected to this line of reasoning on the grounds that, while some operations of the brain were admittedly analogous to running a program, others were not. Mathematics might be programmable, but originality and inventiveness, the identification of relationships that had not been seen before, the discovering of patterns in the apparently patternless, were, almost by definition, unprogrammable. Against this it could be argued – and still is being argued – that the existence of the stored program implies the ability to construct new programs; an instruction can take the form of instructions for the creation of new instructions; a machine can be told to think about its own thoughts. This avenue of exploration is most obviously applicable to what has become known as 'artificial intelligence', which is the subject of a later chapter, and is mentioned here only to make the point that the fact that a machine is programmable does not in principle mean that it is limited to the use of programs supplied from outside.

The computer does, however, have one characteristic which it is far from certain that the brain shares. Just as Babbage's analytical engine concentrated its mathematical operations in its mill, and the Turing machine concentrated them in its head which scanned one symbol at a time, so the computer concentrates its logical operations in a central processor. The processor performs the operations at colossal speed – up to a million times a second against the brain's cycles of about one hundred a second – and it works on groups of either eight or sixteen digits at a time. Nevertheless, this arrangement has the consequence that a computer, however many states it has, can change them only step by step, in stages where each change of state is the result of applying a logical, computed rule. This is, of course, exactly the sort of behaviour that the handling of mathematical processes demands, but it is neither the only way in which it is possible to organize a digital discrete state machine nor, necessarily, the way the brain is organized.

In the early days of computing such arguments could only be hypothetical, in so far as there seemed little prospect of the computer ever attaining the size and complexity of the brain. Since the whole issue centred around the hypothesis that complex information processing in the brain was a result of its capacity to store very elaborate programs, size was all important. The ability of a brain or a computer could be expressed in terms of the number of its memory elements and their capacity, and the memories available to the computer were, at first, severely limited by technological considerations.

Valves were expensive and comparatively large objects which had to be fitted into place in wired circuits and mounted on racks. It was possible to build systems which had switches numbered in hundreds or even thousands, but providing them in millions would require structures the dimensions of which would be measured in miles. It was, however, possible to enlarge a computer's memory (though not the range of its electronic states) by dividing it into two parts, an instantaneous electronic memory, and a less accessible but also less expensive and cumbersome back-up system. The information in valve circuits would be like the stock on display in a shop, available on demand to the customer, the logic circuits of the central processor; the back-up system would be a stockroom or warehouse from which further information could be drawn when new or different supplies were needed.

Most of the early computer projects were principally concerned with finding the right medium for this back-up store. The world's first working computer, built by Dr F C Williams at Manchester University, which ran its first program in June 1948, utilized cathode ray tubes to store the digits as dots of light. Another mechanism that was tried in both Britain and America was the delay line, a tube in which electronic pulses could be kept, bouncing backwards and forwards from one end to the other. The problem was that any medium which stored information in linear form, like a record or tape, would inevitably make it difficult to retrieve quickly, while devices like the cathode ray tube or delay line, which were more immediately accessible, were also complicated and costly. By the early 1950s a reasonable compromise had been achieved in the form of a magnetic drum carrying bands of tiny electromagnets which could be switched on or off by a head travelling up and down from band to band or round the circumference of each band. Equipped with this sort of back-up, a computer was like a student who could remember only a few answers, but had the knack of finding the right place in the textbook and reading it with lightning speed.

But by the time the magnetic drum had established itself as the standard back-up medium in the first generation of commercially available computers, a development had occurred that would, in due course, make the whole question of secondary importance. The invention of the transistor, an electronic switch that is intrinsically simpler, cheaper and more reliable than the valve, transformed the future of the computer. Instead of being a machine that could only be laboriously constructed by hand, like a motor car of the 1890s, it became possible to envisage a time when computers could be turned out like Model T Fords. The transistor not only reduced the need for back-up memories in the long term, it also eliminated the painstaking work of constructing local, fast access memories called magnetic cores on which most machines had been reliant. As things developed, of course, the implications were even greater, for the transistor shrunk from a gadget the size of a cigarette filter which had to be plugged into its place on a circuit board to an infinitesimal fleck in a chip of silicon, which is itself not just a circuit but a complete computer costing no more than the price of a few valves 30 years earlier.

Electronic components with a capacity comparable to that of the brain can now be contained in a medium-sized suitcase and bought for a few hundred thousand dollars; and thus the machine can no longer excuse its limitations on grounds of technological or economic constraint. If the model is right, we have all the ingredients of a fully fledged mechanical brain.

Since we cannot compare the natural and the artificial models directly – knowing that the neurons store digital information and transmit it in the form of pulses is a long way from understanding how the information is encoded in digits or how the transmissions are organized – we can only compare the inputs and outputs, seeking algorithms that enable the computer to do the same things as a brain.

The most important demand that any robot makes on its computer brain is likely to be that it organize and control the movement of a mechanical body, and most of today's robots do not, in fact, ask much more than this. Movement is a process for which it is reasonably easy to find algorithms, and one which fits naturally into the computer's characteristic mode of doing one thing at a time. Some simple examples will show that controlling the movement of a mechanical body comes very naturally to computers.

Let us suppose that a robot brain is asked to control the extension of a robot arm. The position the arm has to reach can be expressed as one series of digits – let us say for simplicity's sake the sequence 1111, and its present position as another, say 0011. Subtracting the second number from the first will give a third number which represents the distance the arm has to travel, and if the arm has a sensor that feeds back information as it moves, then the subtraction will produce a smaller and smaller difference until, when the arm has extended itself to the correct distance, the difference will be nil. Thus, when the operation starts, the logic circuits will perform a subtraction like this to establish that the distance travelled is the equivalent of twelve units:

$$\begin{array}{r} 1111\ (15) \\ -0011\ (\ 3) \\ \hline 1100\ (12) \end{array}$$

When the movement is half completed, the calculation will show that a distance equivalent to six units still has to be travelled.

$$
\begin{array}{r}
1111\ (15) \\
-1001\ (\ 9) \\
\hline
0110\ (\ 6)
\end{array}
$$

When the operation is complete, the two numbers will be the same:

$$
\begin{array}{r}
1111\ (15) \\
-1111\ (15) \\
\hline
0000\ (\ 0)
\end{array}
$$

A program of this kind can obviously be extended to allow for the fact that the arm may be required to slow down as it nears its objective in order to arrive smoothly with no impetus that would cause it to overshoot or knock over the object its end effector has to pick up.

Robot movement can also utilize the computer's logical, as opposed to its mathematical, functions in the same sort of fashion. Imagine now that the need is to ensure that the arm follows a certain path in a straight line, deviating neither to right nor left. The desired path could be defined as one pattern of four digits, register A (a register is the storage device in which the processor holds a pattern of digits while operating on them), and the sensors on the arm could provide information about its actual position in the form of a second pattern, register B.

If each pair of digits in the two registers is then compared (by an 'exclusive or', EXOR, gate) to produce a third pattern, C, that pattern will define the relationship between the required path and the actual path. For example, say that A is 0111 and B (when all is well) 1100. C will be 1011 (EXOR is a logical not an arithmetical function, so the pair of digits on the right, 1 and 1, give 0 not a 'carry'). If the digits of C are listed as C1, C2, C3, and C4, then the instructions in the program will take the logical form. Take no action if and only if:

C1 = 1
C2 = not 1
C3 = 1
C4 = 1

or, in the notation used by logicians, no action $= C_1 \wedge \overline{C}_2 \wedge C_3 \wedge C_4$ (C_1 and not C_2 and C_3 and C_4).

Suppose that the arm strays off course to the left, changing B to 0100, and thus C to 0011; the program would provide for this contingency by an instruction to redirect the movement to the right if and only if:

$$C1 = \text{not } 1$$
$$C2 = \text{not } 1$$
$$C3 = 1$$
$$C4 = 1$$

or, $\overline{C}_1 \wedge \overline{C}_2 \wedge C_3 \wedge C_4$ (not C_1 and not C_2 and C_3 and C_4) = move right.

Programs of this kind may appear to be an extraordinarily elaborate way of telling a robot arm to do things which the human arm does, apparently, without calling on the brain for any significant share of its capacity. But they are exactly the sort of calculations that the computer can repeat hundreds of times a second. It should be added, lest it be thought that this sort of programming is necessary every time a robot takes on a new job, that programs like this are also just the sort of basic skills which can be built into a computer at birth, in its ROM or 'Read-Only Memory' which incorporates all the basic unchanging programs it will require throughout its life. We do not know exactly how the brain carries out similar tasks, but when one watches a robot in action performing complicated and precise movements with apparently effortless fluency, the idea that some very similar process is taking place becomes entirely credible.

Unfortunately, the computer is far less successful when it is required to undertake processes that, unlike movement, have no obvious computational algorithms or which are not intrinsically linear. Understanding visual perceptions, or perceiving patterns of any kind, is not something that is readily reduced to a sequence of calculations. When we see a pattern or spot a relationship between objects, perhaps a curve on the road ahead or a crooked place setting on the dining table, we do not, as far as we can understand or monitor our own mental processes, arrive at an interpretation of what we see by calculation. Even when dealing with something as complex as language, it is notable that a child is reckoned to be able to read at the point when he or she no longer sees the sentence 'the cat sat on the mat' as T-H-E C-A-T S-A-T O-N T-H-E M-A-T, but as a single statement, to be absorbed as an entity, and not as a sequence of letters or syllables. In the same way, when we look at one of Escher's famous puzzle pictures, in which a single image is capable of two quite different interpretations, the brain switches from one to the other in what appears to be a single transformation – the change is like a cut in a film, when the scene changes instantaneously and completely.

A great deal of ingenuity is being applied to finding algorithms which enable computers to recognize patterns – in visual images, language, and other forms – but the going has proved much harder than might have been expected to be the case if the computer was indeed a totally satisfactory model of the brain, and this poses considerable problems for roboticists. For vision, language and reasoning are the skills which robots will require if they are to advance from the status of blind serfs.

But before looking more closely at the kind of limits the computer model may be running up against, and the possible alternatives to it, we must examine the third element of any practical robot: the organization of communications between 'brain' and 'body' without which even the finest brain would be unable to exercise any useful degree of control over a mechanical body.

5. Nervous Systems: The Medium *and* the Message?

What a piece of work is a man!

William Shakespeare, *Hamlet*

The arrival of an electronic brain to partner the available range of mechanical bodies did not in itself make the robot an attainable objective. In order to combine a computer with, say, a mechanical arm to form a functioning whole it is necessary to have some clear idea of how the two should relate to each other. In more general terms, it might be helpful to understand how controlling mechanisms, brains, communicate with the mechanisms that they control, bodies. In the mid-1940s there was no overall theoretical understanding of this relationship and an inventor whose mind turned towards the possibility of incorporating the infant computer brain in a robot body would have found himself in the position of an airline which owned a variety of planes and had at least one volunteer pilot but was handicapped by the discovery that he did not know how to fly and, even worse, that no pilot training manual existed to provide the necessary instructions.

Physiology, which might have been expected to have some hints to offer about how the human brain controlled the human body, was, to continue the analogy, in the position of a highly competent ground crew. It could explain the workings of the various component parts of the 'plane', and knew how to diagnose and remedy a whole range of malfunctions to which they were subject, but it had no advice to offer about flying it. Medical science, in other words, could describe the role of the lungs in respiration or the function of the kidneys and it could diagnose diseases in terms of the symptoms they produced and the failures they caused, but it understood the body's control apparatus, the nervous system, only as a maintenance crew would understand the linkage between throttle and engine or rudder bar and rudder, and even then the understanding was superficial and limited.

It had been known since the early nineteenth century, when the Italian physicist Alessandro Volta noticed that the muscles of a frog's leg twitched if connected to the terminals of his newly invented battery, that the operations of the nervous system involved electricity. But it was another century before the Spanish histologist Ramon Cajal de Santiago traced much of the intricate network and the medical world accepted that its purpose was to conduct pulses of electricity to and from the brain, afferent nerves carrying instructions to the muscles and efferent nerves relaying back the evidence of the senses. To know what the nervous system does, however, is a long way from knowing how it does it; and though it is relatively easy to envisage a network of electric wires simulating the function of nerves in transmitting information, this does not help very much unless it is known what kind of information is to be transmitted, how it should be encoded in pulses and how the brain and central nervous system organize and make sense of it.

Some progress had also been made in describing the basic components of the system. Under a high-powered microscope the physical material of the brain, which is to the naked eye a sludgy jelly reminiscent of a thick pea soup, is revealed as a mass of separate nerve cells or neurons. Even though the size of a single neuron is infinitesimal (ten million of them could fit in a thimble, and the brain contains something of the order of 10^{10} or ten thousand million), the work of the Nobel Prize winners A F Huxley and A L Hodgkin at Cambridge University had provided an explanation of the chemical interactions which caused a neuron to 'fire' or generate a pulse and Warren McCullough and Walter Pitts, working at the University of Illinois, had subsequently shown that neurons generated pulses as a logical function of the firing of other neurons; but no one had any clear idea at all how the firing action of thousands of neurons translated into a 'thought'.

If, in an image which had gained wide currency even in children's books, the nervous system was thought of as a telephone network, with the brain at its centre as the exchange, then science had advanced to the point where it could trace the route of the wires and explain the workings of an individual handset, but the bird's nest of connections inside the brain was labyrinthine and its workings mysterious. To understand a communication system at this level, moreover, reveals nothing about the messages it carries, and to think of the brain as simply a telephone exchange is misleading in that it misses the central point that the brain is also the principal user of the system. Understanding of the nervous system at the more fundamental level of what messages are being passed, what language is being used, what the conversation is about and why, required a completely different approach: one which started by asking the question 'what is it all for?' rather than simply 'what is it?'.

If the would-be roboticist had turned from anatomy to engineering he might have been offered some pointers in this direction. Though, to the layman, even the most advanced machines of the 1940s were still devices that operated under human control – cars needed drivers, planes required pilots – engineering had, even before the advent of the computer, begun to study the problems of designing controlling machines. At least one truly autonomous machine, the steam engine controlled by a governor of the type invented by Watt, had existed for nearly two centuries, and systems like telephone exchanges also operated without human control as did some large scale processes in the chemical industry. There were also many cases in which direct human control had become impractical and had to be implemented by means of servo mechanisms; the helmsman on a large ship, for example, could not move the rudder by his own unaided efforts and a steering engine had to be incorporated to generate the necessary force. The science of control engineering had, therefore, come into being to deal with issues which were, albeit in a very simplified form, similar to those which would be posed by the design of a robot nervous system.

Control engineering had in particular developed two formal concepts which are fundamental. The first of these is the idea that the relationship between controller and controlled takes the form of a 'loop' around which information flows. Control cannot be exercised and sustained simply by a one-way stream of instructions. A drill sergeant, for example, who turned his back on a squad of soldiers and proceeded to bellow commands into space would, within a fairly short space of time, lose track of the squad's whereabouts and find that it had marched off the parade ground and into a ditch or the officers' mess. In order to maintain control he must use his eyes to 'close the loop' and keep himself informed of the squad's position. The information that comes as a result of closing a loop in this way control engineers called 'feedback'. Feedback not only enables the controller to judge when new instructions are called for and what they should be, but it also allows him to check that previous orders have been received, understood and acted upon, that the squad of soldiers, for instance, has not misheard 'left turn' for 'right turn'. Feedback that is used to reduce errors, by providing information that can be used to compare the actual outcome of an instruction with the desired one, is called 'negative feedback'.

The second principle established by control engineering was that the mere existence of a loop is not sufficient to guarantee effective control. The critical factor is the amount of feedback provided and too much can be as damaging as too little. This is an important consideration, even with servo mechanisms where ultimate control remains in human hands. A power steering system in a large vehicle, for instance, must feed back to the driver the right amount of information about the forces produced by the castor action of the wheels in order to provide him with the 'feel' of the vehicle. Too little feedback will result in the steering feeling soft and 'floppy'; too much will transmit every irregularity in the road back as a wrench to the steering wheel. In either case, stable control will be made more difficult.

The amount of feedback becomes even more important when a machine incorporates its own controlling device, which is unlikely to have the flexibility and reserves of judgement available to a human controller. The danger of instability arises when the ability to perceive error and the capacity to correct it are not exactly matched and efforts to correct a mistake in one direction result in over correction, creating very possibly even greater error in the other. A system prone to instability of this kind can very easily degenerate into cycles of ever wilder oscillation.

By reducing these ideas to rigorous mathematical form, control engineering had, by the end of the Second World War, made possible a wide range of autonomous machines and had provided a variety of servo mechanisms. The profession was, however, a long way from the point at which it could contribute much to the design of a nervous system for even a relatively simple robot limb. The instructional and feedback information it dealt with was usually transmitted by mechanical linkages, or their hydraulic or electrical counterparts, which translated movements at one extremity of a loop into corresponding movements at the other by some direct analogue method. The information that was passed thus took the form of force or motion rather than information in the abstract sense which is the medium in which computers operate, and it certainly did not look as if the natural nervous system was a contrivance of levers for converting small movements in the brain into large movements of the muscles or vice versa.

In fact, the differences between the ideas with which control engineering concerned itself and the complexity which, it could be inferred, was involved in a natural nervous system were so much more striking than the parallels that it scarcely occurred to either engineers or medical men that there might be any common ground. Medicine, it is true, was accustomed to using mechanical analogies to illustrate its knowledge to a lay audience – describing the brain as a telephone exchange is but one example of this – but it was not accustomed to the idea that any real insight could be gained from engineering, which dealt with mechanisms of infinitely less subtlety than those in the body. For their part, most engineers as yet had little grasp of the possibilities, let alone the implications, of the new kind of machinery which stored and processed pure information and the idea of mating one of the new electronic brains with a mechanical body was a distant speculation. The gulf between the two areas of science was enormous and to workers on both sides the possibility of trying to bridge it would have appeared remote and, probably, pointless.

To see that this was not so and that, at an abstract level, comparisons could be made involved taking a step backwards away from particulars and looking at both systems in terms of general principles. The man who took that step was Norbert Wiener, and the vista he opened up from the new perspective thus gained was so novel and valuable that it can now be seen to have amounted to a whole new way of looking at both men and machines.

In 1947, when he published *Cybernetics*, the book which both named and inaugurated the new science of studying 'control and communication in men and machines', Norbert Wiener was in his late fifties. Most of his career, which had begun early since he had been a mathematical prodigy in his teens, had been devoted to concerns at least as abstract as those which led Turing to the notion of his universal machine. Though his reputation within the small world of pure mathematics was considerable, he was scarcely known outside it and in retrospect the elevation of a specialist mathematician at the Massachusetts Institute of Technology into the public prophet of the 'information machine' can be seen as signalling the onset of a new age.

Wiener's interest in the possibility of comparing men and machines had first been aroused in the late 1930s when, at a multidisciplinary seminar at Harvard, he first became aware that the gulf between engineering and biology might result more from the lack of any common scientific language than from a total disparity in subject matter, and that the forging of a link might be a proper concern for mathematicians. But the spark that set the tinder of cybernetics ablaze in Wiener's mind was a more specific issue arising out of his passionate anxiety to contribute something to the Allied war effort.

The question with which Wiener became involved in the latter stages of the war was that of devising fire control systems for anti-aircraft guns which would remove the need to rely on the skills of a human gunlayer. The advent of electronic calculating machines suggested a means of doing the necessary computations at high speed, but the difficulties went beyond the need to calculate the amount of 'lead' that had to be allowed when shooting at a fast-moving target. It could be assumed that, while the gunlaying system on the ground was calculating how far the aircraft would travel during the time it took for the shell to reach it, the pilot would be aware that he was going to be shot at and would take evasive action.

Wiener saw that, although individual pilots might take different decisions, whatever manoeuvres the plane went through its behaviour could, from the point of view of the gunlayer, be treated as random within the limits of aerodynamic possibility. To the non-mathematician, this might seem equivalent to saying that any direction in which the gun was pointed was likely to be as good as any other; but Wiener's mathematical work had been much concerned with randomness, and he was able to devise a system which laid the gun in a way which maximized the *probability* of hitting the target. As it happened, the war was over before Wiener's work yielded concrete results; but this was beside the main point, which was that he had realized that practical advantages could be obtained by treating the behaviour of a human being, the pilot, which was in reality the result of free will or predispositions caused by training or character, as if it was mechanical and subject to purely mathematical rules.

The fact that Wiener chose to expand the germ contained in this single example to try and analyse other aspects of human behaviour did not mean, as has often been assumed, that he viewed human beings as machines or thought that all human behaviour could be explained in mechanical terms. He was, on the contrary, deeply opposed by both temperament and conviction to such ideas and saw more clearly than most where they might lead. But, if he saw the threat to human values posed by the new category of information machines, and warned of it long before it was fashionable to do so, he also saw that their development offered an opportunity for gaining further understanding of human beings, particularly those aspects for which, as we have seen, traditional medicine could provide no clear model. His hope was that the knowledge thus gained would help to better the prospects for medical treatment of nervous and mental disease and perhaps, ultimately, enable better artificial limbs to be provided for the disabled. Wiener certainly foresaw that his work would benefit a future science of robotics, but he was quite clear that the motive for building robots should be the benefits that would accrue to all from having a better scientific model of man, not the economic advantages that a mechanical replacement for man might bring to a few.

The science of cybernetics, as originally propounded by Wiener, had four principal themes and it is worth identifying these clearly, if only because they have been partially obscured by subsequent events. The first was the central assertion that there were levels at which men and machines, especially machines which processed information, could validly and profitably be compared. And the level with which cybernetics was primarily concerned was that of constructing mathematical models of the way in which both natural and artificial systems handled information in the expectation that this would provide new insights into concepts in philosophy and psychology.

The second and third themes proposed two such models, one suggesting a yardstick by which information of all kinds could be measured, and another showing how ideas developed by control engineering could be applied to the human system.

The fourth theme was essentially a speculation: that the way in which the brain processed information was a consequence of its structure and that understanding of its workings would be gained by building artificial structures which had similar characteristics.

Before looking in detail at the first and fourth themes, which are obviously closely interrelated, it is worth examining in more detail the second and third in order to understand their importance and relevance to robotics.

In setting out his second theme, Wiener drew largely on the work of Claude Shannon who, after his research on Boolean logic circuits described in Chapter Four, had taken a post with Bell Telephone Laboratories where he had developed a precise theory of information. Shannon's theory was based on the notion that information was essentially a commodity by which uncertainty is reduced according to statistical rules. A simple example is the familiar panel game of 'Twenty Questions' or 'Animal, Vegetable or Mineral', in which a team seek to identify an object which they only know to fall into one of the three categories by asking a limited number of 'yes' or 'no' questions. The skill lies in posing questions which reduce the range of uncertainty by the greatest possible degree. Thus, if the object is 'animal', the first question may well be 'Is it human?' and, if the answer is 'yes', this might be followed by 'Is he or she still alive?' Shannon suggested that the reply given to each question of this kind constituted an irreducible minimum of information, one 'bit', and that any systematic arrangement of information would always form a pattern in which each bit reduced uncertainty by fifty percent. A pack of cards provides a good illustration of the point. If one wishes to establish, by a process of elimination, which card someone is holding out of the range of 52 possiblities this can always be achieved with six questions. Supposing that B is holding the six of spades, A's interrogation might take the following form:

A: Is it red?

B: No (Therefore it must be a spade or a club.)

A: Is it a spade?

B: Yes

A: Is it seven or greater?

B: No(Therefore it must be a spade between the ace and the six.)

A: Is it four or greater?

B: Yes

A: Is it five or greater?

B: Yes

A: Is it five?

B: No (Therefore it must be the six of spades.)

A moment's thought will show that, using this method, six questions will in fact suffice to identify any card in a pack of up to 64 cards, where each card is different and the pack forms some distinct hierarchy. This is equivalent to the statement that, in binary arithmetic, six digits suffice to denote any number between 0 and 63. Shannon's basic unit, the bit, the 0 or 1 stored by a single electronic switch has of course become familiar as the measure of the capacity of a computer's memory. Computers normally deal with information in blocks, made up of eight bits, known as bytes, or, latterly, in larger units of 16 or even 32 bits. Each address in the memory serves to identify the location of one byte, which is taken in and out of store and handled by the central processor as a single unit. If we apply the formula, described in Chapter Four, which states that a binary number with x digits can convey 2^x messages, it is obvious that an eight-bit byte can serve to represent a number up to 2^8 or 256: say one character out of the whole range which makes up a keyboard. It could also be used to denote the position of a robot articulation or the extension of a joint to an accuracy of one part in 256. The use of 16 bits would, in theory at least, reduce the uncertainty to one part in 65,536 (2^{16})!

Shannon's work had another aspect which concerned interference or 'noise' in a communications system – a subject clearly of great importance to his employers. But, to a communications engineer or cybernetician, noise is not simply a crackle on the line, it can take the form of any interference which is random in occurrence, a misprint in a newspaper or the misreading of a dot for a dash in Morse code. One way of thinking of Wiener's original problem about anti-aircraft artillery is to consider evasive action by the pilot as noise which would affect the communication of the plane's motion to the gun on the ground and, as in this case, the chances of noise occurring in any system have to be estimated in terms of probability. Shannon showed how, in any given system, one could calculate the chances of a single bit, transmitted as a 1 or yes, being received as a 0 or no, or vice versa.

Though Shannon's theory provided a model which has been perfectly adequate for analysing the kind of information that contemporary industrial robots require, it has not provided the universal model for which Wiener foresaw a need. The difficulty arises when one tries to analyse a highly complex system of communication, such as a language, in statistical terms. Unfortunately, grammatical structures cannot be fitted into a model based on the Shannon theory; moreover, in all except the simplest instances, there will be a level of 'meaning' which depends upon the context. The difficulty comes about because, in dealing with language, the human mind is able to make assumptions that go beyond what is contained in the communication in rather the same way as we automatically assume that a tree, the top of which we can see above the roof of a house in the foreground, has a trunk, or that the river on one side of a bridge is the same river as that on the other. Much of the richness of language results from this depth of meaning which makes devices such as simile and metaphor possible, so that we can use forms of words which are statistically identical, such as 'Joe is flying high in the office this week' and 'The plane is flying high in the sky today' while remaining confident that a listener will distinguish the difference in the meaning of the phrase 'flying high'.

For similar reasons, although it would be possible to analyse the play in any game of cards with a system which used the six-bit descriptions based on the Shannon principle, this would not work if the game was one in which aces were 'wild' because although the number of cards has not increased, some of them have a quality, the ability to take on a different meaning in different contexts, which is outside the statistical grammar of the pack. In an attempt to construct models of language – and other kinds of information which have this multi-layered structure – in mathematical terms, later workers adopted algebraic rather than statistical methods; though, as we shall see, this approach has not been totally successful in reducing everyday, natural language to a model which can be understood by machines of the kind presently in use.

Wiener's third theme, the application of control engineering principles to natural systems, is the one that has won the widest acceptance and has had the most obvious influence upon robotics. In retrospect, indeed, it seems surprising that an idea so obviously 'right' should have appeared so revolutionary at the time. The fact that the human function of control involves loops and feedback is clear from such an everyday phenomenon as giddiness. When someone spins round and round and becomes giddy they lurch and stagger because the organs in the inner ear which measure the body's orientation relative to the vertical are disturbed and fail to feed back accurate information to the brain and, in the absence of this feedback, the brain simply 'doesn't know' which are the appropriate controls to apply.

Using the mathematical methods employed by control engineering, Wiener and his colleague Arturo Rosenblueth were able to explain a much more serious clinical condition, the disease ataxia. This affliction causes a sufferer to lose the ability to position a limb accurately as is required, for example, if one wishes to pick up a pencil from a table. Instead of the hand reaching out smoothly to a position where the fingers can grasp the pencil, the patient's lower arm oscillates back and forth, causing the hand first to overshoot and then undershoot its objective. Rather than being due to a malfunction in the muscles of the affected limb, ataxia was, Wiener and Rosenblueth showed, caused by a failure in the loop formed by the brain, arm, and eye. By measuring the rate of oscillation in the affected limb, they proved that the problem lay in the mechanisms which detected the position of the arm muscles. These failed to transmit accurate information to the brain and, as a result, the brain issued a succession of instructions and counterinstructions which led to the observed instability.

One of Wiener's early followers, the British biologist W Ross Ashby, took matters a stage further when he showed that the superstable motor functions of the body such as breathing, metabolism, and heartbeat maintain their regularity, not because of some astonishing process of control engineering performed by the brain – a sort of lifelong balancing act – but because the control system itself enjoys the property of homeostasis or inherent stability.

This was one of the first triumphs of cybernetics in that the examples Ross Ashby used revealed that parallels could be found between, apparently, enormously complicated systems in nature and machines of a relatively trivial kind. But, despite this initial success, the approach suggested by Wiener's third theme could not, it transpired, be pursued very much further. The difficulty was that, while the mathematical methods involved could be used to account for linear processes where each effect has a single cause, they did not help in the analysis of more complicated, non-linear processes such as occurred in the brain. For it was already known that neurons are not connected in chain-like systems, where the effect of a pulse at one extremity would be a ripple of pulses running along the line, but in intricate nets in which a pulse at one point can result in wholly unpredictable behaviour at another. Information in the brain, in fact, flows through an inconceivably complex series of loops and feedback arrangements the behaviour of which can only be explained with the rather different methods of automata theory. These methods are examined in Chapter Seven, but for the moment it will suffice to note that the difference lies in the fact that, whereas in classical control theory it is possible to think of any combination of changes in a system happening in sequence – one thing leading to another in a well-defined chain of cause and effect – automata theory can deal with systems in which one effect may have many causes and where change may be discontinuous and disconcertingly dramatic.

Wiener's original work, however, together with that of Ross Ashby and others, laid the foundations on which most existing systems of robot control are based. Indeed, they gave rise to some of the very first robots, the 'cybernetic tortoises' produced by another British scientist, Grey Walter, in the 1950s. These small electrically propelled vehicles were equipped with devices which measured the reserves of power stored in their batteries and sensors which enabled them, when supplies ran low, to seek out and 'plug in' to a recharging apparatus. One point about the 'tortoises', which were one of the star attractions at the Festival of Britain in 1952, was that they achieved the apparently sophisticated task of satisfying their own appetites with a minimal 'brain', and many observers were led to suppose that they were much 'cleverer' than was, in fact, the case.

In robotics, however, the main importance of rigorous models of loops of information and feedback systems is that they make it possible to design machines which combine precision with stability. A simple example will illustrate the point. In Chapter Three one common form of robot drive, the stepper motor, was described in detail and it was explained that its mode of operation, revolving by one step for each pulse of current fed into it, makes it particularly suitable for computer control. Unfortunately, stepper motors are not perfect and some system of feedback is required to monitor their performance. Let us suppose that a robot elbow is required to rotate through an arc of ten degrees for some purpose, and the computer program calculates that this movement involves feeding 30 pulses to the stepper motor which provides the drive. It is entirely possible that the motor will miss or gain a step and move by 29 or 31 steps instead of the prescribed 30. How does the robot's computer detect and rectify the error?

One simple form of feedback sensor is a device called a shaft encoder, a transparent disc mounted on the axle of the motor and turning with it and so arranged that light shining through it from one side is registered by a battery of photo-electric cells on the other. The disc carries a series of black blotches, each of which is a segment of one of a series of concentric rings, and which cut off the light reaching one or more of the photo-electric cells as the disc rotates (see diagram overleaf); a pattern of ten rings read off by ten cells will provide 1024 (2^{10}) combinations of binary digits enabling the cells to distinguish its position to an accuracy of one part in 1024, or about one third of a degree of rotation. In our example, therefore, a shaft encoder would register a missed or gained step on the part of the motor as an error and feed this back to the computer which would then supply a pulse to the motor to rectify the problem and the shaft encoder could verify that this had put matters right.

It can be seen that it is essential that the instructions and the feedback distinguish positions with a similar degree of precision. If, for example, the shaft encoder could detect errors of one part in a thousand, whereas the motor operated in steps equal to one part in five hundred, the feedback would always be in danger of discovering errors the drive was incapable of rectifying. The human system is prone to problems of this kind as well, even when it is not suffering from any disorder. It is a common experience, for example, when trying to thread a fine needle, to find that the hand holding the thread begins to waver more and more wildly the harder one concentrates on the task, and that the only solution is to relax and start again, not an expedient that is readily programmed into a robot. One answer to this threat of overcorrection is to reduce errors in stages, aiming not to solve the whole problem in one go, but to

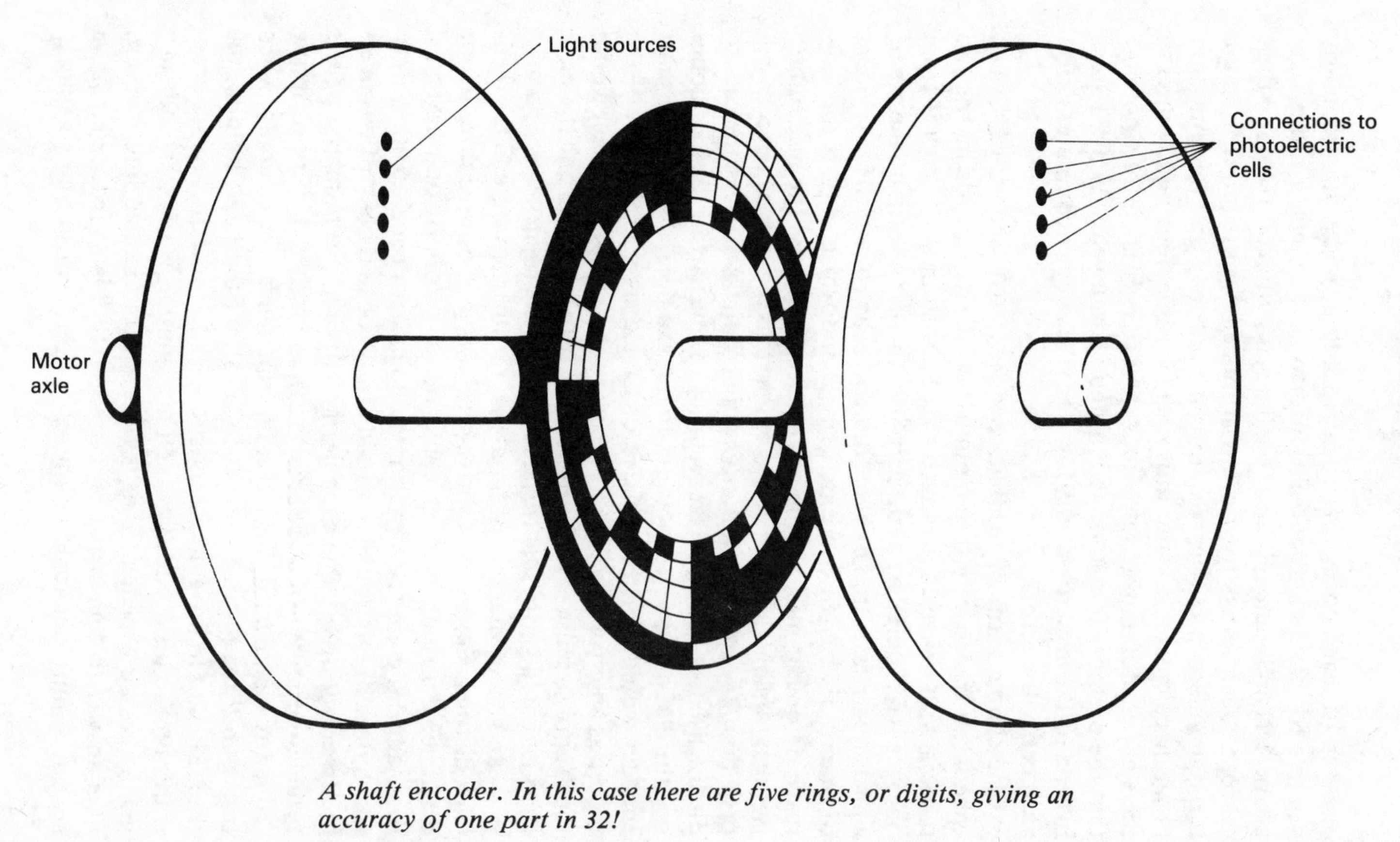

A shaft encoder. In this case there are five rings, or digits, giving an accuracy of one part in 32!

narrow it down in rather the same way as the range of cards was narrowed down in the example on page 96, until the error is within the required tolerances.

Many robot control systems are, of course, a good deal more complicated than the example of the shaft encoder might suggest. The gap between the actual position of a joint and the desired position can, for instance, be measured as a difference in electrical potential; or strain gauges can be used to judge the leverage that is being exerted on a joint or the pressure that an end effector is applying to an object in its grasp. The choice of sensors employed in a particular robot will be dictated, in part, by the kind of drives that are involved and by the degree of accuracy which is required. One of the principal criteria of robot performance is what is known as 'repeatability', that is, the accuracy with which the arm repeats a movement once it has been learned. Some applications obviously demand a higher degree of repeatability than others – spot welding robots, for example, may manage quite adequately if they can position their welding guns to within an eighth or even a quarter of an inch (the sheets of metal that are to be welded often have 'dents' stamped into them to guide the clamps of the gun into position). Smaller robots, designed for intricate assembly tasks perhaps, can position their end effectors with far greater precision, to within a few thousandths of an inch or, in the case of some of the very latest machines, three or four microns (millionths of a metre), a feat which makes even the most dextrous brain surgeon seem positively clumsy by comparison.

It must, however, be remembered that what is involved, in the case of robots, is not the fine coordination of hand and eye which a human being relies upon but a blind, mindless process of repetition which depends for its precision entirely on proprioreceptive senses (the senses that keep a check on the position of the robot's own moving parts). When a robot is 'taught' a sequence of movements for the first time it learns the position each joint is required to take up at every stage by noting, in the form of computer programs, the information fed back by its various sensors.

To understand how this teaching process is carried out in practice, we might imagine that it is necessary to teach a robot arm to move from point A to point B. The human operator will normally use a so-called 'teach pad', rather like a more elaborate version of the radio control sets used by model enthusiasts to guide their boats or planes, which is linked to the robot by a cable. Assuming that the arm has six degrees of freedom, the pad will carry six separate switches which the operator can use to control the six joints, steering them by using the switches to flex or straighten a joint or extend or contract the arm. The first step will be to steer the arm into position A; this can be done in any old fashion, but the important thing to establish is that the position is attained with absolute accuracy. Once he is certain that the arm is in exactly the position he requires, the operator will press a 'teach' button which causes the computer to record the digital information from the encoders or other sensors that define position A. In less sophisticated robots, which employ what is known as 'point-to-point' control, the teacher would then steer the arm to position B and instruct the computer to record the details of the new position. When the robot was required to repeat the movement it would simply calculate the most direct route from A to B and follow it.

But most contemporary robots employ the more complex techniques of 'dynamic control', which allow the teacher to instruct the machine in the exact route that the arm is to follow and the speed at which it is to move. Dynamic control is essential if the arm is to avoid obstacles or, say, to follow a curved or angular seam with a welding gun; it also enables the teacher to ensure that the end effector moves smoothly from one position to another and arrives at its destination with a minimum of impetus. Where a very intricate pattern of movement has to be followed, when spray painting a convoluted shape, for example, a slightly different teaching method may be used in which the teacher takes hold of the robot's end effector and steers it through the required pattern, perhaps in 'slow motion', while the computer brain assiduously catalogues the changing information from the sensors.

Dynamic control demands the provision of separate feedback loops and control systems for each variable factor. One sensor, for instance, will feed back information about the leverage being exerted on the joint by gravity, and this will vary according to the load carried and to the degree of extension of the arm. Others will monitor the speed of movement and the rate of change in that speed to ensure that the arm arrives at its destination with a minimum of impetus, a calculation that will vary according to the size of the movement and the inertia involved. Each sensor will feed its information back into the control system at the appropriate place, completing a smaller loop within the major one formed by the overall position control and the position sensors.

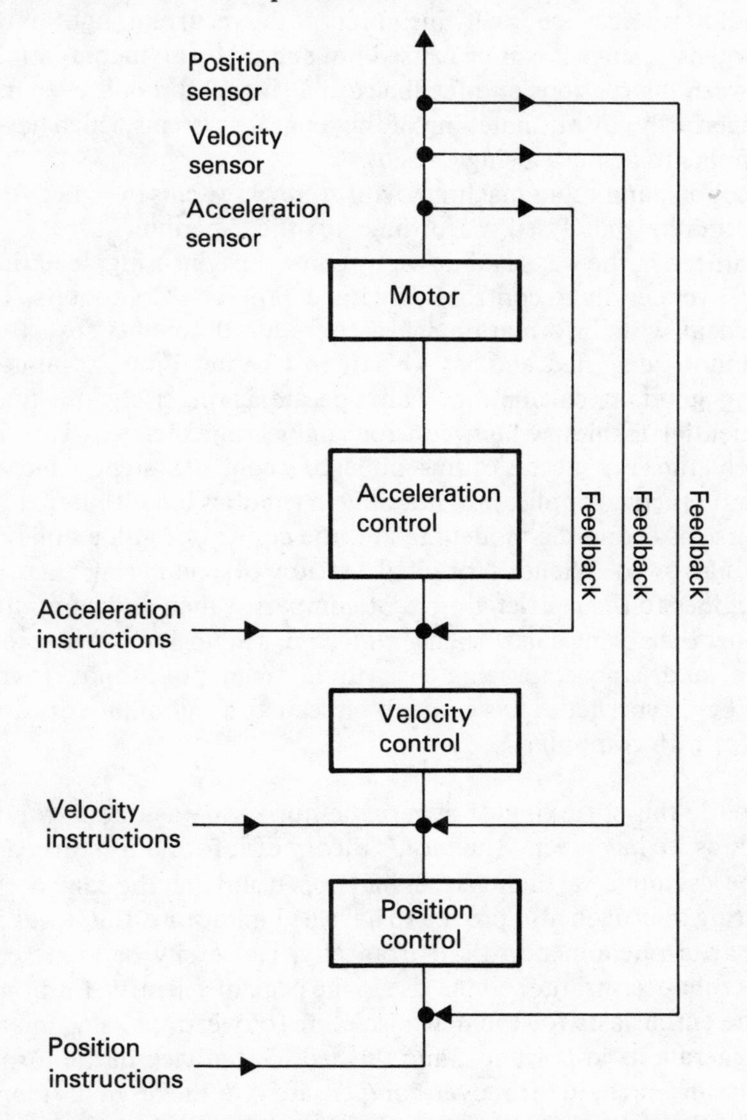

Diagram illustrating the arrangement of control circuits and feedback loops required to achieve dynamic control over a single articulation in a robot arm

It can be appreciated that designing systems which provide controls of this kind for each of six articulations, and which superimpose an overall control of the whole arm, requires complex and exact mathematical models and sensors of several different kinds. It should also be evident why instability is the recurrent nightmare of roboticists, since it can be caused not simply by an incompatibility between instructions and feedback in a single loop but, even more dangerously, by mismatching of the separate systems which have to combine to achieve a single result.

Robots, and other machines with control systems designed to use information analysed according to the Shannon theory and organized in the ways laid down by control engineering, lend themselves very easily to control by digital computers. Computers, after all, deal with information which is reduced to 'bits' exactly as Shannon suggested and, as we saw in Chapter Four, computers, being good at calculation, can operate in precisely the linear, sequential fashion which control engineering dictates. Unfortunately, the very success of this model of a control system, which has found practical application not only in robotics but also in fields as diverse as economic modelling and the control of guided missiles, has narrowed Wiener's original version of comparing men and machines to the restricted field of comparing men and computers. Rather than being a new science which opened up new vistas across a wide landscape embracing everything from philosophy to psychology, cybernetics has come to be seen as a technique for 'doing things with computers'.

In part, this shrinking of cybernetic horizons was due to the fact that, as we have seen, Shannon's theory of information turned out to be less universal than Wiener had hoped and that the control engineering approach also proved to have its limitations. But it was also due to a phenomenon that, ironically, can easily be analysed in cybernetic terms: the sudden rise to the peak of scientific fashion led some enthusiasts for the new discipline to overstate its claims and exaggerate its importance, and this led to feedback in the form of criticism which, in turn, over-compensated for the original error. In particular, cybernetics advanced claims in fields such as management and social psychology which were not really justified. This laid it open to attack by people who, although they were not remotely qualified to assess its mathematical methods, found it easy to ridicule their application in the social sciences. Social cybernetics became, briefly, as fashionable as social Darwinism had once been and the reaction was all the swifter and more effective because it was relatively easy to misinterpret, or even misrepresent, Wiener's argument for *comparing* men and machines at *some* levels as a program for *replacing* men with machines at *all* levels. It was, perhaps, unfortunate that over-enthusiasm on the part of his disciples was mirrored by traits in Wiener's own character which led him to savour his sudden elevation to the status of prophet with a certain ponderous relish.

One consequence of cybernetics' dizzy rise to popularity, and the sinister, anti-human associations it acquired, was the creation of a whole subspecies of science fiction characters, half robot and half living, tagged with names like 'cybermen' and 'cybernauts'. Sadly, the popular image these beings represented has proved more influential than Wiener's own attempt to set the record straight with his book, *The Human Uses of Human Beings*, published in 1950.

For the future of robotics, a serious result of the fall from grace suffered by cybernetic ideas was the comparative lack of attention which was paid to Wiener's fourth theme, an omission that was encouraged by the all-conquering advance of the computer.

When *Cybernetics* was launched upon the world in 1947, the computer was only just undergoing its birth pangs, but Wiener was already well aware that the arrangement of electronic switches proposed by von Neumann was only one of a whole range of possible 'architectures'. Indeed, Wiener's whole approach to the possibility of information processing machines was quite different to that of von Neumann and the other computer pioneers. For the inventors of the computer, the fact that parallels could be drawn between their valve circuits and nerve cells was an intriguing, but ultimately irrelevant, curiosity. If it meant anything, it was that nature and engineering had arrived at the same solution to the problem of storing digital information by a process evolutionists call convergence, that is, the two systems were similar not because one had been inherited or copied from the other, but because there was, perhaps, only one solution possible.

Certainly when it came to planning the structure of the computer the mathematicians and engineers did not look to nature for a model; instead the arrangement adopted was dictated by the idea of the computer as a programmed machine that would resolve mathematical and logical problems. Von Neumann's design for such a machine has remained unchanged in fundamentals throughout the whole process of technological refinement and growth in capacity and power which followed. The modern microprocessor is as much a 'von Neumann machine' as its first, primitive forbears. Indeed, the von Neumann model has been so successful that it has tempted many to assume that all information processing machines must be preprogrammed to perform logical operations in organized sequence. The corollary of this is the belief that intelligence, the most vivid expression of the brain's processing powers, is a product not of the structure of the brain but of the complexity of the programs it contains.

On this argument, it follows that the structure of the brain is a matter of subsidiary importance, for the secrets of intelligence lie in the programs that are run, not in the machines which run them, and even if a computer is not like a brain this need not prevent it from achieving similar results. All that is required is to find the algorithms which reproduce those results when programmed into a computer. It is only within the past few years that the problem of finding algorithms which enable computers to deal with complicated information processing tasks like understanding natural language or interpreting visual information (problems which are examined in detail in Chapters Eight and Nine) has led to renewed interest in Wiener's fourth theme, the proposal that machines with brain-like structures might have inherent, brain-like qualities which are not dependent upon preprogramming. The work of putting this proposition to the test goes under the heading of 'modelling neural nets'.

The three concluding chapters of this book are devoted to the work that has been done on the modelling of neural nets and the expectation that, as Wiener hoped, alternatives to the architecture of the von Neumann computer may allow us, in true cybernetic fashion, to draw parallels between man-made machines and that most mysterious of all mechanisms, the human brain. It may be, therefore, that the most important of Wiener's contributions to our understanding of men and machines is still to be realized; in the meantime, this chapter has demonstrated that cybernetics was the key which allowed a start to be made on combining two kinds of machine, the traditional, 'mechanical' kind and the new, information processing kind, in order to create a third kind of machine, the robot.

6. A World Fit for Robots

Eyeless in Gaza, at the mill with slaves.

John Milton, *Samson Agonistes*

It has often been pointed out by biologists that little can be gained by studying the behaviour of an isolated individual of a highly social species such as ants, for only an entire community of ants behaves in a way that is truly 'ant-like'. Much the same argument applies to any examination of the role of the contemporary industrial robot. A robot in the unautomated environment of an earlier industrial age (an eighteenth century iron foundry, say, or a nineteenth century locomotive works, or even a mid-twentieth century aircraft factory) would have been an ingenious but useless curiosity. An ant without an anthill. For the technologies used in robotics and the applications of robots in the workplace are so intertwined with the wider field of automation that it is impractical to consider the role of robots in isolation from the development of automation in general.

A lively debate is currently underway over the future relationship between robotics and automation. In the view of one camp, robots are likely to prove a temporary, short-term expedient which will become unnecessary as true automation advances. Opposing this argument are those who claim that robots have the potential to outpace any conceivable developments in other areas and to climb the ladder of skills so rapidly that other techniques of automation will be relegated to subsidiary roles. Much of the dispute is, at base, a question of terminology since, although robots are dependent at present upon the technologies of automation in general, ideas and techniques which are in one way or another robotic increasingly pervade the whole field. In order to understand the terms of the discussion and its implications for the future of robotics, it is necessary to start by examining the nature and aspirations of automation itself.

The word was first coined by a works manager at the Ford Motor Company in the 1950s to describe a process that was just gathering speed and which, it was becoming clear, was different in kind from the mechanization of industry which had been underway for a century or more. Mechanization could be defined as the effort to equip a human labour force with better and more powerful tools. By the middle of the twentieth century, many of these tools had become automatic, it is true, but only within fairly narrow limits. Most machines still depended upon the skills of a human operator, and manufacturing was generally divided into distinct stages defined by the machines and skills involved. The invention of the moving production line had transformed some assembly work into a continuous process, but it was still largely a means of breaking the assembly process down into separate stages and bringing the work to the human worker, reversing the traditional assumption that the workpiece remained stationary while the worker and his tools moved about.

Automation involved, from the start, two novel ideas which are distinct, though related. The first was the substitution of mechanical for human control over a single machine or process. Perhaps the best known example of this is the numerically controlled or NC (and, later, the computer numerically controlled or CNC) machine tool. The point is that, since the repetitive machining of identical components involves a highly skilled machine operator in repeating the same process over and over again, it is possible to program a machine's cutting head to repeat a sequence of movements and thus automate the whole task. A machine automated in this way differs from something like an automatic bottling machine, say, in that it is relatively easy to reprogram it to perform different jobs without major engineering modifications, and it thus retains the flexibility of an operator-controlled machine while gaining the labour-saving economy of an automatic one.

The second, and in many ways the more important, aspect of automation was the integration of separate machines into combinations which could perform a whole sequence of tasks automatically. Thus, for example, the essential elements of a number of different machine tools can be combined in a so-called 'transfer machine' which can carry out the whole variety of machining, drilling, routing, and finishing processes necessary to transform a raw casting into a finished component. The gains in efficiency made possible by this, second, kind of automation can clearly be seen in a quite different field, that of book production.

The manufacture of books has involved the use of automatic machines for the past century at least, but until very recently the processes of printing sheets, folding them into 'signatures', binding the signatures together to form a 'book block' and inserting it into a 'case' had been treated as entirely separate operations involving different kinds of machinery and skills and often carried out in different factories. With automation came machines capable of transforming blank paper into finished, bound and jacketed books in a single 'pass', only the prefabricated cases and the jackets having to be produced separately. Skilled labour is still required to 'make ready' the machinery for each job, but the unskilled work of stacking the sheets, transporting them from place to place, and feeding them into each separate process has been eliminated with gains in both economy and speed.

The creation of integrated systems of machinery to perform a whole series of functions relies, in turn, upon reliable means of moving things about while retaining precise knowledge of and control over their position and orientation. For once a component or workpiece is dumped in a bin or stacked up in a corner on an unorganized pile, the intervention of a human worker is necessary to get the automated process restarted. Traditional manufacturing is a little like a game of football in which the ball goes out of play every time a player kicks, requiring the game to be restarted. The aim of automation is to keep the ball in play by planning longer and longer sequences of mechanized movement, ideally continuing from the moment of kick-off, when the raw materials are fed in, to the goal of final assembly.

Both aspects of automation are, like the robot, products of the 'information revolution' of the past thirty years. Traditional engineering employed automatic machines the functions of which were determined by their construction; changing the function, if it was possible at all, involved changing the components and much labour with gauges and spanners. With the arrival of 'information machines' in general, and of course the computer in particular, it was possible to change a machine's function by changing its program of instructions, in a way that had hitherto only been feasible with a very limited number of machines such as the Jacquard loom from which Charles Babbage had borrowed his punched card input system more than a century before. Enabling machines to utilize information and integrating them into combinations required that they be able to communicate with one another in a fashion which had, again, scarcely been possible before the advent of information processing and the concepts which cybernetics developed and refined. One pioneer of automation, the British industrialist Sir Leon Bagrit, even suggested that a more appropriate description of the whole undertaking would be 'cybernation'.

Automation thus shares many of the ideas and techniques which are involved in robotics. Instructions have to be transmitted in digital form and formulated as algorithms on which computers can operate; feedback has to be provided at every stage and, where separate processes are integrated into a single system, a hierarchy of 'intelligent' control must be installed. In a sense, automation has involved building machines 'in the image' of a whole variety of professional men. The numerical controls of a machine tool could be seen as a model of the professional skills of the human operator. In the oil and chemical industries, computer systems substitute mechanical judgement for the experience of the engineers who previously read off the information about flows, temperatures, and reactions and adjusted the controls accordingly. In some cases automation has even produced systems which are designed in the image of the manager: computers can, for example, organize the work of a car factory where bodies may have to be painted in different colours or be fitted with different engines or trim, in order to ensure that the right bits arrive at the right place at the right time and that inventory is kept down to an economical minimum. In all these instances, automation is performing in ways which could be described as robotical.

The role in automation of robots proper, in the strict sense of machines which possess articulated arms, is, however, curiously equivocal. Though they are sometimes seen as the most complex and elaborate manifestations of automation, they are often employed only in order to make good its inadequacies or to fill awkward gaps left by other systems. It has even been suggested that robots, because they are designed to mimic human actions and attributes, are the enemies of real automation rather than its allies. In order to see why robots should be perceived this way, it is necessary to understand that, in the view of many of its proponents, automation only comes into its own when it goes beyond simply substituting a mechanical hand for a human hand on the tool or control button and changes the whole way in which things are done. The essential point of this argument can be seen in a very elementary example, that of securing packages. To design a machine which reproduced the human action of wrapping a cord around a package, tightening and then knotting it, would be a complicated task; especially if it was required to deal reliably with packages of different sizes and shapes. The banding machine, flinging a tensioned metal strip around a parcel and sealing the joint automatically, achieves the same result, a securely sealed package, but using materials that a human worker would find awkward and dangerous and a method he could not possibly emulate.

This principle of rethinking an industrial process completely in order to automate it effectively goes well beyond the machinery and methods employed. In many cases, it has been found necessary to redesign the product as well. Take, for example, the electrical circuit. The natural way for human labour to wire up a circuit is by connecting wires to components which are prepositioned on the appliance or circuit board. The printed circuit, which turns the whole concept on its head by creating the 'wiring' first and then plugging in components as required, is ideally suited to automation but more or less impractical for human hands to perform. It is a classic instance of a technique which is intrinsically 'mechanical', proving more efficient and economical than any possible refinement of the 'human-based' methods which preceded it.

Many of the applications in which robots are currently employed arise because what might be seen as imperfect or incomplete automation has left a niche which can be filled only by human labour, or by a machine which mimics its actions. The whole field of 'parts transfer' (moving things from one place to another), in which robots are widely used, can be seen as a consequence of imperfect automation and will, it is suggested, vanish when more elaborate, more adaptable, and better controlled systems of conveyors, equipped with automatic gates to direct traffic, come on to the scene. Another common robot application is the loading and unloading of machine tools and the need for this comes about, it could be argued, as a result of incomplete automation: machine tools require a man-like machine to service them only because they are still designed to be serviced by men. Future generations of machine tools may incorporate loading and unloading arrangements that are themselves automatic. The same goes for diecasting and moulding machines and other specialist installations which are currently 'customers' for robot labour.

Those who support the view that robots are likely to be a temporary, compromise solution to problems which full automation will eventually solve by other means can point to the fact that the present generation of robots are in any case dependent upon, if not actually parasites of, larger automated systems. Blind, insensate robots could not, for instance, move components about or service other machines were it not for the fact that automation makes it possible to deliver items in predetermined positions and orientations and provides an electronic communications network which keeps the robots informed about where everything is and what is happening.

Even in cases where the robot arm has proved to offer the best possible answer to a particular problem, it is possible to envisage a time when it will become a mere limb or appendage of a larger system rather than a separate machine in its own right. In the areas of welding and paint spraying, for instance, where an articulated arm provides an easily reprogrammable device for reaching over or around obstacles or into awkward corners and is particularly valuable because it can 'track' a moving production line, the trend is towards specialist robots designed with only one function in mind. To the extent that such machines are not adaptable to different purposes they could be considered something less than real robots. It is, moreover, a fine point as to whether the arms are the 'real' robots or simply the obedient slaves of the larger machines which transport and position the workpiece.

In many ways this whole controversy, when applied to the kinds of robots currently in use, is somewhat artificial, hinging as it does upon the formal definition of a robot as a machine with an articulated arm. This is especially evident when very simple and comparatively unsophisticated robots, often limited in terms of degrees of freedom and precision of control, are compared with, say, the more advanced vision systems which are currently being developed and which, because they are designed only as sensors and interpreters rather than operators, do not qualify for robot status. It would be difficult to argue that a machine capable of rather elementary dexterity was a more accurate image of man than one which could observe and 'understand' its environment. Such a suggestion comes dangerously close to claiming that arm-wrestling is an inherently more advanced human activity than astronomy or aesthetics.

The most likely outcome seems to be that robots will become more disparate, more specialized and more integrated into larger systems but that, at the same time, automation as a whole will become increasingly dependent upon techniques that are, in the widest sense, robotical. The result will be large combinations of automated equipment which, taken as a whole, perform tasks which demand robotic qualities but in which it will not be possible to point to one particular component and identify it as a robot. In this view, industry will not be a home for a reinvention of man, but only for those aspects of man which are industrially valuable and can be readily integrated into a new kind of machinery, the all-embracing automatic production system.

There is, however, quite another view of the future of the industrial robot which it is possible to adopt. The starting point in this case is the fact that automation has thus far been restricted to relatively large scale industry and to processes which demand limited degrees of intelligence, adaptability, and dexterity. Automation has, for example, made little progress in assembly work, which is usually too 'fiddly' for mechanical methods. Nor has it, as yet, been adopted in businesses with short production runs or rapidly changing products, for the equipment for large scale, 'hard' automation tends to be expensive and specialized. It is true that new approaches to what is called 'flexible manufacturing' are tackling some of these limitations; but in the real world the change from old methods to new is a gradual one involving compromise, makeshift arrangements, and much managerial caution. For the time being at least, the disincentives to automate are considerable. The prospect of rapid change in the future tends to discourage precipitate investment in the present, and automation is still very much an 'all or nothing' business, for the implications of automating one aspect of an enterprise tend to ramify into all the others. There is little point in clearing one bottleneck in the main street of production if the result is a new traffic jam elsewhere.

The suggestion is that considerations of this kind will always act as a brake on hard automation in many areas of industry, but that the robot will soon provide an attractive alternative. The picture that is drawn by adherents of this school is of a future in which all except the largest scale industry retains a recognizable shape, continuing to use the kind of methods and machinery currently in use but replacing human labour with robot labour at an ever increasing pace as robots become more adept, flexible, and intelligent.

It must be said that this line of thought involves assumptions about progress in robotics that many would consider optimistic, if not unrealistic. In particular, it assumes the relatively imminent arrival of extremely sophisticated vision systems together with a degree of intelligence that can 'make sense' of what is seen. As will be shown in Chapters Eight and Nine, it begins to look as if the 'thinking' aspect of automation will have to develop beyond the confines of conventional computer methods. Some idea of the kind of problems, often fairly elementary but nonetheless serious, that arise from attempts to replace human labour directly with robot labour in the absence of this sort of advance can be gleaned from a single example which arose in recent experience. A robot was installed to take the place of an operator at a machine tool station, and it was only when the new system was ready to go that the snag emerged. The tool involved produced long coils of metal swarf, a fact that had been overlooked in assessing the suitability of the task for a robot, simply because the human action of sweeping a pile of swarf on to the floor from time to time appeared so trivial. It was only when the machine tool was engulfed in a mass of serpentine metal that it was realized that swarf disposal was a problem with which the robot had no means of coping. In the event, the difficulties were overcome, but only by means of installing a pneumatic system to vacuum the swarf away, a solution that involved substituting a raw material that produced a dust-like waste that flowed freely though the suction pipe.

Some robot inadequacies can be overcome by 'cheating'. Even the most elaborate vision systems currently available, for example, have great difficulty in interpreting quite simple three dimensional images, a matter of some importance in an application like arc welding where the robot end effector may have to maintain a precise distance from the welding seam while tracing a curved or angular path. One way round this problem currently being investigated is the use of structured light, light which is projected through a grid in order to throw a pattern of light and dark stripes; this emphasizes the contours of an object in a fashion which lends itself to digital interpretation. In other cases when only the detection of the presence or absence of an object is required, ultrasound devices may offer a substitute for vision.

But the use of special lighting or of methods which demand the re-design of the product, will clearly limit the robot's utility as a replacement for a human worker. Moreover, robots dependent upon techniques like this demand lengthy setting-up procedures, coordination with other automatic systems, and, very likely, a considerable period of 'debugging' in each new application. They are a long way from achieving the versatility of even an unskilled human worker and would certainly be more trouble than they were worth in most small businesses. The same seems likely to hold good for many of the more advanced vision and sensor systems that are at present being developed, for they will 'recognize' only a very limited range of simple objects and only after rather extensive 'training' or 'teaching' processes. For large scale industry, with long production runs of identical products, this may not represent a major handicap, but it will certainly prevent robots gaining a foothold at the lower end of the industrial scale.

What may be more imminent than the robot jack of all trades, indeed is already appearing, is the small, relatively speedy and dextrous robot which can undertake simple assembly work. Interestingly, the size and shape of many of these machines are dictated by the need for them to work alongside human beings, occupying the same floor space and having the same radius of action. Robots of this sort are still, of course, dependent on a production line which delivers parts in a predetermined fashion – they work blindly and can at present perform only the most elementary tasks. An alternative to robots like these, designed to work side by side with human beings on a 'traditional' assembly line, is the robot which incorporates its own 'environment' – a small, self-contained 'world' which is carefully designed to meet robot requirements. Such 'worlds', 'flexible manufacturing systems' as they are called, take the form of a small work-bench about the size of an average table covered by an overhead gantry from which the robot arm is suspended so that it can move about and reach every area of the bench with equal ease. The bench can be fitted with various pieces of equipment – a jig to hold components in the course of assembly, a tool rack where the arm can store and change its range of end effectors, and racks to hold supplies of parts which can be replenished from outside as they become exhausted. The point is, of course, that within this microcosm of the factory floor everything the robot needs can be positioned with total accuracy and, once it has been taught a process, the arm can move about its little world with complete confidence. In order to ensure that the arm keeps its bearings with total accuracy some fixed reference point may be built into the work surface so that, from time to time, the robot can recalibrate the position of its end effector.

A system of this kind is an excellent illustration of the way in which robots and automation in general are becoming increasingly interdependent, for it cannot be said that in such a case the robot arm has an existence separate from that of the system as a whole. The robot, if robot it be, is designed specifically to work within its purpose-built world; one could not take it out and expect it to play a useful role in a different context. Nor would a human worker find a little world of the robot arm a comfortable one in which to work. The system is not simply a workplace for robots, in the sense that an airplane's cockpit is a workplace for a pilot or a locomotive cab a workplace for a train driver. Instead, the little world of the work-bench and the robot which operates in it are matched to each other like a snail to its shell and it is no more possible to detach one from the other than it is reasonable to extract a snail from its shell and expect it to stay alive and well.

It is likely not only that assembly robots will gain in skills, but also that the design of products will start to take the limitations imposed by robot assembly into consideration – it is, for example, relatively easy for a robot to assemble components on a 'pancake' basis, which involves no more than placing things one on top of another and securing the whole collection with a single screw. Another aspect of matching the product as well as the workplace to robot labour involves the comparatively new field of computer-aided design (CAD); it is already being proposed that the information generated by a CAD system – the sizes, dimensions and tolerances of components, for example, might be transmitted directly to the memories of the robots that will be responsible for production, bypassing the need for the robots to be trained for a new task by human programmers.

Though all these developments point towards a widening of the robot's industrial repertoire, none of them suggest any real break in its dependence upon an automated environment, indeed, rather the reverse. As new sensors, interpretive techniques, and control systems become available, they are likely to take the form of self-contained 'faculties' which can be added on to existing robots or inserted elsewhere in the overall anatomy of automation. It is less likely that they will all be incorporated in elaborate (and expensive) multi-purpose robots. Thus, though it seems likely that there will be plenty of robot arms lifting, probing, and groping in the factory of the future, the most realistic prognosis seems to be that automation and robotics will grow closer and closer together and, in some cases, meld into a single discipline, making the most of the robot's ability to reproduce human skills where these are needed, but also capitalizing on automation's talent for finding new and better ways of doing things which depart altogether from the idea of a factory based around human labour.

The kind of developments that may be expected in industry can already be seen in the more familiar circumstances of the home. Since the 'domestic robot' has been one of the staples of science fiction, whose robots rarely dirty their hands on the factory floor but are often to be found doing the housework or serving the cocktails, it may be legitimate to use the home as an illustration of what we mean when we talk of automation and robotics becoming a single subject.

One of the striking failures of early science fiction was its inability to postulate a future in which all technology, rather than just the most obviously sensational aspects, had advanced. Thus robots of the interwar years were commonly shown wielding household appliances such as carpet sweepers and flat irons that would, in the event, become outmoded long before the robot came on the scene. It is this kind of technological tunnel vision that the supporters of general purpose automation warn against when they argue that the industrial robots of today will be superseded by machines of a quite different kind rather than by improved robots. It is extremely probable, for example, that the need for ironing of any kind will have been eliminated by advances in fabric manufacture before anyone has produced a robot which could safely be entrusted with a steam iron and the week's laundry.

It is, on the other hand, perfectly possible that some sort of floor cleaning robot might appear on the market in the near future; such things are already in use in large supermarkets. Optimism should, however, be tempered by the knowledge that, at the present stage of development, use of such a machine would involve the laying of magnetic guidance tracks under the carpet, and the fixing of all the furniture in set positions; nor could it be relied upon to discriminate between, say, the area of carpet where the guest of honour spilled the *boeuf strogonoff* the night before and the corner of the room which never gets dirty at all – it can only apply a standard process to all areas impartially. Since the machine would be more like an automated vacuum cleaner than a mechanical dustpan and brush, it would be unlikely to sweep underneath the sideboard or pick up the toys the kids had left lying about. In order for it to be effective, it would probably be necessary to round off the corners of every room and there would certainly have to be a lengthy training process, repeated whenever you decided to rearrange the furniture. On the whole, the average household is likely to be better off with an ordinary vacuum cleaner.

In fact, it is probable that the development of air filters and dust extractors would resolve many of the problems of household cleaning far more effectively than any conceivable robot. For the arrival of the microprocessor has already made the automated house a practical proposition. You can, if you wish, install automatic systems to draw the curtains, open the door, monitor the temperature room by room, do the laundry, dispose of the garbage, or record TV programs that might otherwise be missed, making use of 'off-the-shelf' silicon chip intelligence and existing machinery. Automation may, in short, leave little for a robot to do in the way of domestic chores. But to suggest that this means that robotics has nothing to offer is to miss the point. For many of the devices currently available owe something to robotic principles, and it is likely that the debt will increase in the case of their more advanced successors. The misapprehension science fiction suffered from was that all robotic functions would be concentrated in a single machine with two legs, a deferential manner, and preferably, a tail coat.

To illustrate this point it is only necessary to extrapolate a few contemporary examples a few years into the future. Already appliances like washing machines, dishwashers, and food processors have been developed to the point at which, if it seemed useful, a microprocessor or two could provide them complete autonomy. Most modern central heating and air conditioning systems can work for weeks or months on end without human control. Using the techniques employed in the WISARD project (see Chapter Ten) it would be perfectly feasible to depute to a machine the job of recognizing who was at the front door and admitting them if they were family or friends. And if you are alone in the house and no one comes a-calling it is always possible to get out your chess playing computer and challenge it to a game. All these machines could be considered as robots to a greater or lesser extent, in that they are capable of perception or judgement at some level and carry out functions that were previously the sole province of human beings.

In conclusion, it seems probable that the definition of the robot will broaden to include a whole variety of machines which, in one way or another, reproduce some aspect of human behaviour. Many of these machines will be extremely limited in themselves but the availability of cheap, readily adaptable silicon 'mini-brains' will allow them to be interconnected and combined to perform a whole range of tasks. The resulting machinery will not amount to a surrogate human being, but it may be a very adequate replacement for the rather limited range of human talents which are currently demanded by industrial processes.

7. Automata: Finding the Mind in the Machine

It is strange that a little mud
Should echo with sounds, syllables and letters,
Should rise up and call a mountain Popocatepetl
And a green-leafed wood Oleander.

W J Turner, 'Talking with Soldiers'

The preceding six chapters have attempted to provide an overview of the 'state of the art' in robotics, as represented by the contemporary industrial robot. We have described the technologies, especially the information processing and control technologies, employed with the intention of showing what robots currently can and cannot do and of giving some insight into the factors which determine their range of abilities. The conclusion we draw from this survey is that the prospects of the industrial robot, and therefore its future impact on industry, depend crucially upon the chances of developing better robot senses and real robot intelligence. Unless robots can be equipped to see, hear and understand at a level which at least approximates human perception, comprehension and common sense, the expected refinements of their anatomy and physique (better hands, faster movement, greater precision, mobility and so forth) will result in only marginally increased utility, and that only within the context of general industrial automation.

The same argument applies with even greater force if we turn to the possibility of robotics providing a model of man that is useful or illuminating in any serious way, for the aspects of ourselves which we continue to find most puzzling and elusive are the processes by which we think and perceive.

The latter part of this book will, therefore, concentrate on robot brains. This is not to dismiss mechanical bodies as uninteresting or insignificant. Much of the robot's potential interest as a model of man is indeed due to the fact that it constitutes an artificial version of the continuum of brain and body; but it does seem probable that the key to understanding the relationship between the two is to be found in the brain. No study of the engineering or control systems of the body appears likely, by itself, to resolve the mysteries of the brain, but it does seem possible that, if we gain a full understanding of the brain, many of the remaining puzzles about the body will be resolved.

Furthermore, when we talk of 'understanding the brain', we are really thinking of the mind – that is we are concerned with what the brain does rather than what it is. The one may, of course, depend upon the other, but since the construction of any mechanism that physically resembles the brain is quite out of the question, the only practical course open to roboticists is to study and attempt to reproduce the inputs the brain receives, the outputs it generates and the processes by which one leads to the other. It is, after all, reasonable to suppose that if a machine could be devised which worked like a brain and in conjunction with a body, then it might have, or acquire, a 'mind' and that, in creating it, we would have resolved the philosophical riddle posed by the existence of a 'ghost in the machine'.

Any machine which could be said to have a 'mind' would naturally be different in kind as well as scale from those everyday bits of hardware which we commonly think of as machines, and it follows that even discussing the possibility may demand an approach or descriptive system which is different from that which we use to discuss sewing machines or airplanes. This need for a new perspective was in fact something that began to emerge at an early stage in the development of the computer and of information science. For although the first information machines were the result of a painstaking extension of existing electronic technology, and the information they processed was largely formulated in terms of well-known mathematical and logical processes which human beings had mastered, it became clear that autonomous machines which controlled themselves by processing information were a radically new breed, the behaviour of which could not be satisfactorily encompassed by any established theoretical system.

The importance of theory to any branch of engineering is that it enables scientists to predict the behaviour of hypothetical machines as well as describing that of existing ones. In the first half of the nineteenth century, for example, the study of electricity and magnetism, and their exploitation in engineering, were hampered by a lack of any general understanding of the phenomena. Victorian laboratories contained many electrical machines, and magnetism had been known and used for centuries, but there was no established body of natural law which allowed researchers to say, in advance, how an electrical or magnetic system would behave. That could only be discovered by a process of trial and error. The knowledge accumulated through that process eventually allowed the British physicist James Clerk Maxwell to formulate his famous equations which showed that there were not two phenomena but one, electromagnetism, and which set out the laws governing its behaviour. Henceforward, engineers had a set of 'rules of the game' which enabled them, in principle, to predict the behaviour of any electromagnetic system.

By the mid-1950s the infant science of information technology was very much in need of some similar, overall theory which could provide a framework within which the behaviour of all kinds of information machines, the hypothetical as well as the actual, could be predicted. Claude Shannon, as we saw in Chapter Five, provided part of the answer by defining information itself, but there was still no method of analysing all the ways in which it might be mechanically processed. The issue came to a head at a conference called by Shannon himself and John McCarthy (who will be introduced in the next chapter as the founder of the field now known as 'artificial intelligence') at Princeton University in 1956. The topics raised included questions such as: Is there any limit to what can be achieved by logic circuits of the kind Shannon had devised in 1938? Are there any problems which can be shown to be insoluble by any means, even by the use of machines not yet invented? These are questions which can, clearly, only be tackled in the context of some theory which governs all information machines, just as Maxwell's equations govern all electrical machines. The title of the conference was 'Automata Studies' and it marked the point at which the word 'automaton' was generally adopted to describe all automatic and autonomous machines and the birth of a theory, automata theory, which can be used to describe and predict their behaviour.

Lest the reader fear that we are about to plunge into an intimidatingly technical subject, it must be said immediately that automata theory is neither abstruse nor necessarily fearsomely mathematical. On the contrary, it enables us to talk simply and usefully about machinery which would be very difficult to understand if discussed in other, more familiar terms. Furthermore, because it is an all-embracing theory, it can be applied to very familiar, everyday machines which we do not normally dignify with the title of 'automata', as well as to more complex machines.

One other great advantage of automata theory as a way of looking at the new kind of machinery is that it enables us to sidestep the taboos which surround any comparison between human intelligence and 'mechanical intelligence'. For while it is perfectly possible, and often highly illuminating, to think of human behaviour in terms of automata theory, it does not involve thinking of people as machines. The theory is much more a way of thinking about the processing of information, something that, unless one resorts to metaphysical explanations, human beings undeniably do, than about the nuts and bolts of the machines themselves. The design of a theoretical automaton may indeed enable one to produce an actual automaton, but it does not follow, as will be seen, that the result is mechanical in the traditional sense of behaving only in ways which are predictable and determined in advance. Indeed, one of the great benefits of automata theory is that it shows that the question of determinism, which lies at the heart of our suspicions that the robot might somehow reveal our own freedom to be an illusion, is relative and not absolute.

The first thing to establish is exactly what we mean by the word 'automaton'. This is particularly important since it carries with it a good deal of semantic supercargo acquired in everyday usage. Most dictionaries offer two definitions which are, on the face of it, contradictory. The first definition, 'machinery with a concealed mechanism causing its movement', encompasses the sense in which the word is used to describe the clockwork automata which were mentioned in Chapter One. Used in this way, the word implies an element of ingenuity. But when it is applied to human beings, as in the second dictionary definition, it means a person who is 'acting automatically or without intelligence'. A mechanical automaton is, it seems, in some way cleverer than other machines while a human automaton is behaving more stupidly than other people. The apparent contradiction is, of course, explained by the implicit assumption that if the 'concealed mechanism' of the mechanical automaton is examined the reasons for its 'movement' will become obvious. It follows that when a human being behaves like an automaton he acts as if he, too, contained some concealed mechanism which makes his actions predictable and deterministic. The ideas of the behaviourist school of psychology rest upon the assumption that people do indeed contain a series of such mechanisms which determine their response to any stimulus and which can only be modified by conditioning. The behaviourist man is, therefore, an automaton in the dictionary sense. The automaton of the roboticist or information scientist is, however, a little more complicated than that.

The central notion of the automaton, as we shall use the word, is of a mechanism with 'a set of internal states'. The states are, in fact, a more general and useful version of the dictionary's 'concealed mechanism'. The idea of a mechanical 'state' was introduced to describe the position of the components of a calculator or computer in Chapter Four; most importantly, it will be recollected, the state of an electronic switch, which can only be either on or off, represents one bit of digital information in a computer's memory. The term was also used in a rather different, though related, way in discussing the Turing machine. The head of the machine might change its 'configuration', or state, in a manner laid down by its instructions when it scanned a particular symbol, and the state it was in determined what action it would take when it scanned another symbol.

An automaton has internal states of this second, Turing machine, kind. The concept is very close to that of a person's 'state of mind' and the analogy can be used to define exactly what we mean by an automaton. Take the case of an individual whose 'state' is one of anger. That state is likely to be the consequence of something that has happened to him in the past, in terms of automata theory a past 'input', and it will determine his reaction, or 'output', when he is faced with a new 'input'. Thus anger caused by the discovery that the car tyre has a puncture might cause a parent to react more vehemently than would otherwise be the case when a child misbehaves, whereas a state of exhilaration resulting from having been given a raise by the boss might result in the same misdemeanour being treated with indulgence. An automaton is, therefore, a machine which has a set of possible internal states, which are concealed (that is, there is no way of finding out what its internal state is simply by looking at it), which are the consequences of past inputs, and which determine its future outputs.

The number of states which an automaton has will always be

finite and, for the purposes of discussing machines that can actually be built, they will have to be considered as digital and discrete; that is to say the theory can only talk about automata that are in a definite state and has nothing to say about automata that are in the process of changing from one state to another, something that is assumed to happen instantaneously. The function of the theory is to define the rules which govern the change from one state to another and to formulate these rules in terms of the inputs the automaton receives and the outputs it generates. In other words, if we know the past inputs to a machine, automata theory enables us to predict its actions and reactions in any relevant circumstances and to do so by enumerating the changes of state inside the machine.

The inputs and outputs of an automaton may be either actions or information, indeed the automaton itself may exist only as information. Since the theory can provide a set of rules for any possible automaton, it follows that the rules define the machine, or *are*, in some sense, the machine. We met this paradoxical idea of a set of rules constituting a machine in the case of the Turing machine and the computer. The point was, it will be recollected, that the computer is not a machine, like a washing machine or a blender, constructed to do one particular thing, but a universal machine that can do anything for which it is possible to provide an algorithm or set of instructions. The instructions are, therefore, a set of rules which tell the computer what kind of machine it should be in order to tackle a particular task. In the same way the rules which automata theory uses to describe an automaton will themselves fully define the machine, even if it is never constructed.

In point of fact, any automaton which can be described according to the theoretical rules can be built, either as a purely mechanical device and/or as a program for a computer, which may, if the rules describe an automaton that must perceive things in the physical world or take action in the physical sense, be equipped with a mechanical body. Naturally, a computer itself is an automaton within the formal definition, albeit one that can only deal with inputs and outputs which take the form of information, and the same applies to a Turing machine. In fact, as we shall shortly see, a computer program is a set of rules defining an automaton – the computer which contains that program. But it will be helpful to begin the exploration of automata theory by examining a very commonplace automaton, and one which has an entirely mechanical incarnation, the cigarette vending machine.

Smokers will be aware that cigarette machines come in two basic kinds. In the case of the first, simplest type, there is a separate slot in which to insert a coin for each brand of cigarettes, and if a coin is placed in the Marlboro slot then only the drawer containing Marlboros can be opened. The machine is, in fact, a collection of separate machines each selling one brand of cigarette. A theoretician would call an automaton of this kind 'trivial', and with some justice. It can accept only one input and produces only one output. In fact, it has only one working state, that in which it produces an output when given an input; it makes no choices among alternatives for it can only do one thing. Its instructions can be contained in a single sentence: 'If a coin is inserted then release one pack of cigarettes.' It is worth noting in passing that the triviality of this kind of mechanism has not prevented behaviourist theories, which essentially reduce human behaviour to a series of separate mechanisms each producing a fixed output in response to a given input, gaining wide acceptance.

However, even cigarette machines are capable of rather greater subtlety. Consider now the more sophisticated machine in which the purchaser first inserts a coin, and only then chooses the brand he requires by pushing a button. For simplicity's sake we will deal with a machine that contains only two kinds of cigarettes, Brand A and Brand B. Such a machine clearly has two possible states: the state it is in before a coin is inserted (State 1), when no amount of button pushing will cause it to yield up a packet of cigarettes, and the state it is in after a coin has been inserted, when pushing a particular button will produce a pack of the specified brand (State 2). The program which its construction incorporates has to cope with three possible inputs (coin inserted, button A pressed and button B pressed) and provide for two possible outputs (delivering a pack of Brand A or a pack of Brand B). The program also has to take account of the fact that it must not disgorge any cigarettes until it has received payment and that it must never release more than one pack at a time.

It is possible to describe the rules the machine follows by means of a truth table, where the horizontal axis shows the inputs and the vertical axis the states of the machine. In each case, 'looking up' the appropriate square in the grid will tell us two things about the machine: what it will do next, if anything, and what its next state will be.

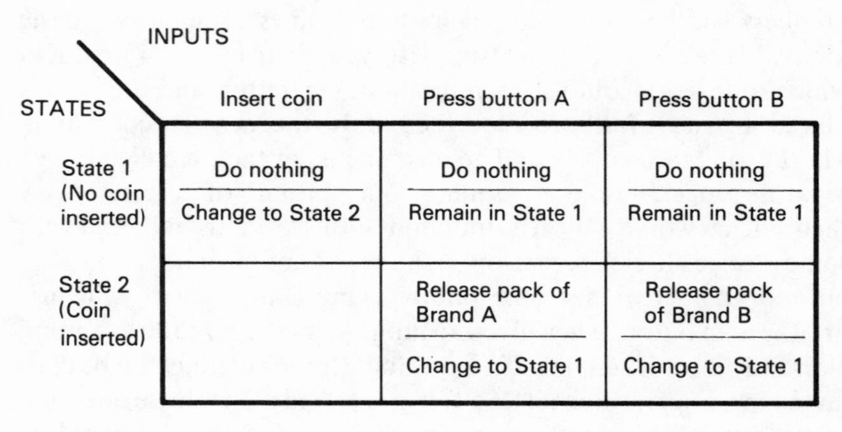

STATES \ INPUTS	Insert coin	Press button A	Press button B
State 1 (No coin inserted)	Do nothing / Change to State 2	Do nothing / Remain in State 1	Do nothing / Remain in State 1
State 2 (Coin inserted)		Release pack of Brand A / Change to State 1	Release pack of Brand B / Change to State 1

Truth table for a cigarette machine

One immediate consequence of setting the rules out in this form is that it becomes obvious, from the existence of a blank in the lower left hand square, that the instructions have failed to take account of the possibility that two coins will be inserted one after the other. It would be possible to get round this problem in two ways: either its occurrence could be forestalled by constructing the mechanism in such a way that it could not accept a second coin until it had 'sold' the first pack of cigarettes; or the square could be filled with the instruction, 'Do nothing/Remain in State 2', in which case after a button had been pressed and a pack delivered the machine would follow instructions and return to State 1, in which state it would detect the second coin and switch to State 2, ready to deliver a second pack.

The behaviour of this cigarette vending automaton can be described graphically by the use of what is called a 'state diagram' which shows the 'state structure' of the automaton. Each state is represented by a circle and the transitions between states by arrows which are labelled with the inputs that bring them about and the outputs that accompany them. The state diagram for the cigarette machine will look like this:

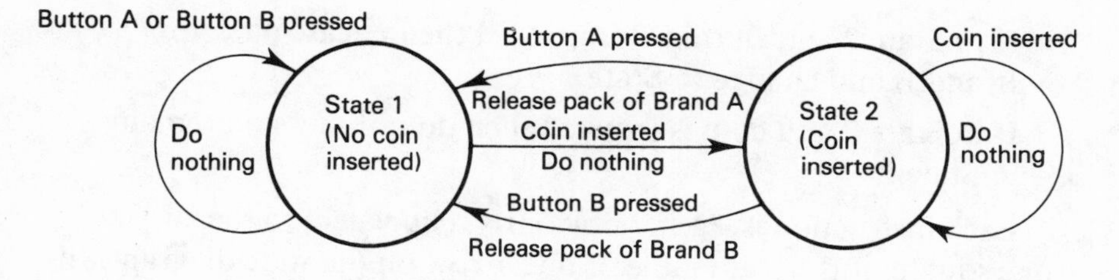

State diagram for a cigarette machine

Note especially the arrows which emerge from each state and circle back to it, showing the circumstances under which the machine should stay in its present state despite an input. This illustrates the general point that 'any input' (and 'no input') are valid inputs which may bring about a change of state just as 'no output' is also a valid output that may accompany a change of state.

It is relatively easy to see ways in which purely mechanical components can be arranged in order to produce even the second, more complicated cigarette automaton. It is only necessary to arrange matters so that the insertion of a coin somehow 'cocks' the two buttons so that they release a pack of cigarettes when pressed. But it is also easy to see how instructions for such a machine might be put in a form which an elementary computer could handle. For one of the fundamental qualities of the computer (indeed a quality without which it is not a computer at all) is the ability to handle logical operations which take the form 'if . . . then . . . '. It will be remembered that Claude Shannon had shown that any arrangement of his logical gates which incorporated feedback (that is, any system in which the result of one logical operation becomes a factor in the next one) had this property. Looking at both the truth table and the state diagram for the cigarette machine, it becomes clear that its behaviour can be set out as a series of 'if . . . then . . . ' instructions or, to put it more plainly, as a computer program. The exact instructions needed are:

If in State 1 and Button A or Button B pressed then do nothing and remain in State 1

If in State 1 and coin inserted then do nothing and change to State 2

If in State 2 and Button A is pressed then release pack of Brand A and change to State 1

If in State 2 and Button B is pressed then release pack of Brand B and change to State 1

If in State 2 and coin is inserted then do nothing and remain in State 2

Each instruction is, rather obviously, equivalent to the answer in one square of the truth table or one arrow on the state diagram. In practice it will be convenient to combine all the instructions relating to being in State 1 and all those relating to being in State 2 into a single instruction. Each numbered instruction in the computer program can then be looked upon as a means of enabling the computer to 'remember' everything it needs to know at that point, or in that state. Since the machine is, by definition, reading only one instruction at a time it does not need to be told what state it is in, and the program consists of a series of orders in the form: 'If (input) then (output) and (next instruction).' Moreover, since the word 'if', when used in a program, generally implies 'and only if' it is often superfluous to tell the machine what to do when there is no input. (In practice, it is also often possible to ignore the need to specify the next state, since a computer is generally instructed to move from one instruction to the next in a numbered sequence unless told otherwise.)

This may seem a rather elaborate way of describing the behaviour of a cigarette machine, especially since one cannot avoid having the sense that the release of a pack of cigarettes is a consequence only of the physical action of pressing a button rather than of the machine 'remembering' that it is in a particular state. We now turn to a machine where a state structure more obviously serves the purpose of allowing it to 'remember' something it needs to know. The example chosen is a rather simple-minded robot.

Suppose that a robot has been installed at the end of a conveyor belt and is required to sort bolts that are delivered to it into two categories, the 'satisfactory' ones, which it must stack in a box, and the 'substandard' ones which it must discard on to the floor. The robot will obviously have to be equipped with a sensor that enables it to distinguish between the satisfactory and the substandard bolts as they appear at the end of the conveyor. In practice, it is likely that this sensing device, if it performs a visual inspection, will itself be an automaton that is a good deal more complex than the robot, but for the moment we can assume that as each bolt arrives in a position to be picked up the robot receives an electronic message telling it whether the bolt is 'satisfactory' or 'substandard' and that one or other of these messages, or 'no input', constitutes its input and that its outputs are 'pick', 'stack', 'discard' or 'do nothing'. Unlike the cigarette machine, this robot requires three internal states because, once the bolt is picked up, there is no further input (like the button pressing on the cigarette machine) to tell it what output is required; it must be able to remember whether it is dealing with a satisfactory or a substandard bolt and determine its own output accordingly.

It should be noted in passing that it might seem that the problem could be sidestepped by providing the robot with two programs, one of which it ran in response to a 'satisfactory' input and one of which it ran when the input was 'substandard'. We would then have, in essence, two trivial automata which could be brought into use as required. But a moment's thought will show that this is not the answer, for a third automaton will then be required to choose which of the other two is to be employed each time a component has to be handled.

The three states required by the robot could be labelled as follows:

State 1 Awaiting a bolt

State 2 Holding a satisfactory bolt

State 3 Holding a substandard bolt

The truth table defining its complete instructions is shown overleaf.

Again the problem emerges of the automaton receiving further input while it is (both literally and theoretically) in no state to do anything about it. Providing the sensor continues sending the appropriate message until the robot has returned to State 1, this eventuality is easily taken care of by filling in the blank spaces with the instruction to dispose of the bolt it is holding in the appropriate way and return to State 1. The state diagram for the robot is shown beneath the truth table overleaf and it is relatively easy to see how it would translate into a program of 'if...then...' instructions.

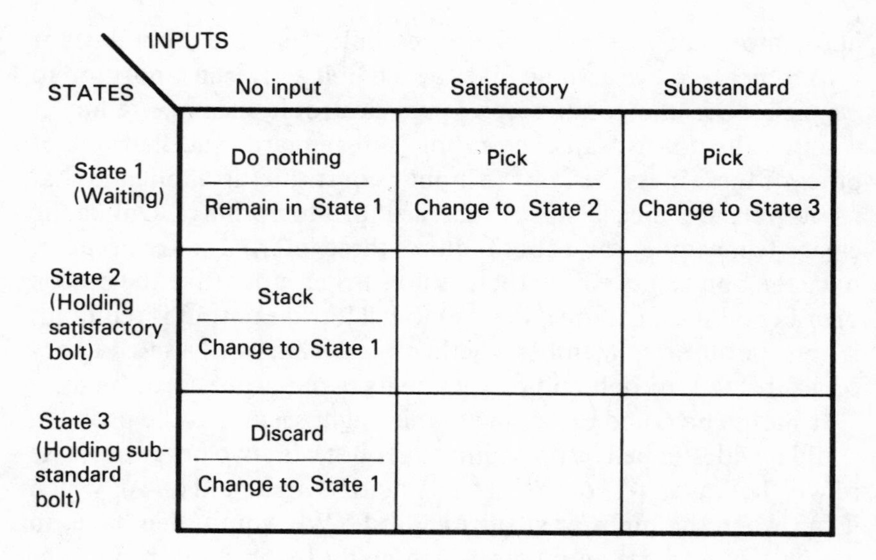

STATES \ INPUTS	No input	Satisfactory	Substandard
State 1 (Waiting)	Do nothing / Remain in State 1	Pick / Change to State 2	Pick / Change to State 3
State 2 (Holding satisfactory bolt)	Stack / Change to State 1		
State 3 (Holding sub-standard bolt)	Discard / Change to State 1		

Truth table for a bolt-sorting robot

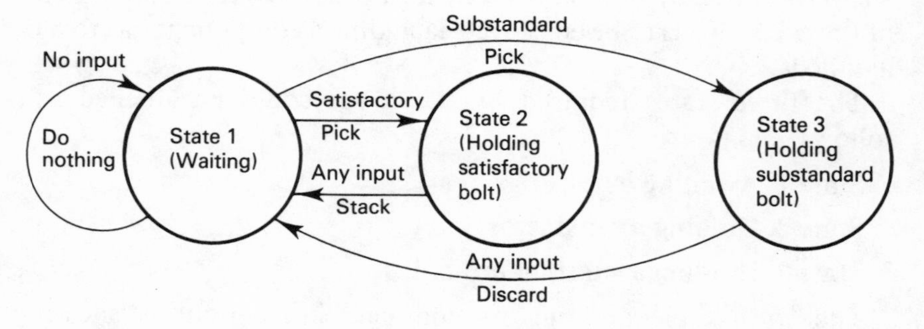

State diagram for a bolt-sorting robot

Imagine now that the owner of this robot finds that business is bad and would be greatly improved if only his robot could be persuaded to treat every other substandard bolt as if it were satisfactory and put it in the box rather than on the floor. With a little ingenuity it is possible to construct a state diagram (shown opposite) that defines a robot which would do exactly that.

With this state structure, the robot will treat the first substandard bolt as before and discard it, but it then goes into a 'cheating' state, State 4, which causes it to treat the next substandard bolt it

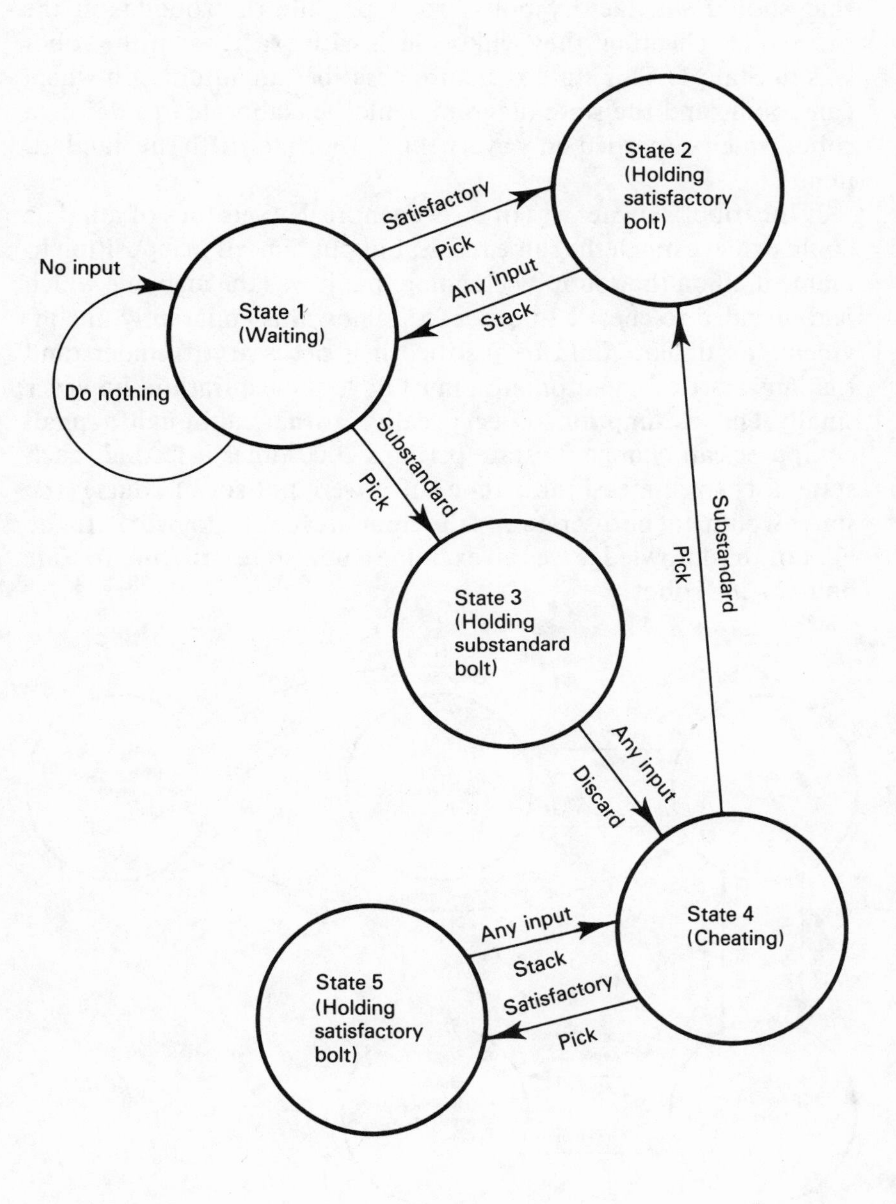

State diagram for a 'cheating' robot

encounters as if it were satisfactory; the existence of State 5 ensures that should satisfactory bolts crop up while the robot is in the process of 'cheating' they will be dealt with exactly as if the robot was in State 1. This state structure describes an automaton which can count, and the state diagram could be elaborated to define a robot which discarded only every third, fourth or fifth substandard item.

If the robot's owner was of an even more devious turn of mind he could achieve much the same result, but put himself in a position to blame it all on the robot by claiming that it was the machine which had 'decided to cheat'. In order to see how this apparently unconvincing argument could be justified it is necessary to understand that any state of an automaton must have some duration, however small. This assumption reflects reality in that, although a small computer can change its state perhaps 2000 times a second, each state lasts for a measurable time – if it were not so, of course, the states would not be discrete and the computer could not work. In the light of this knowledge we can examine a new state structure for our bolt-sorting robot.

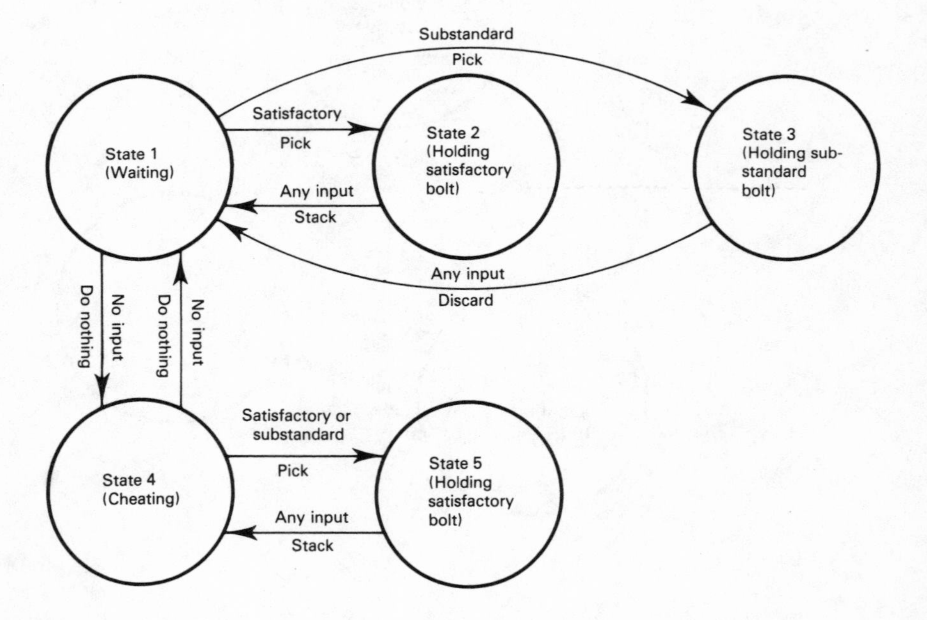

State diagram for an unpredictable robot

In the absence of any input telling the robot that a bolt has arrived, it shuttles back and forth between States 1 and 4 a thousand times a second. When a bolt does appear, it is therefore a matter of chance whether the robot is in State 1 or State 4. If it is in State 1 it will behave as before, sorting the satisfactory bolts from the substandard ones, but if it is in State 4, it will put all the bolts in the satisfactory category. In principle its behaviour is entirely determined, but in practice one might well be tempted to say that it was making up its own mind about whether it cheated or played it straight. Those who are statistically minded might, after observing its behaviour for a while, note that cheating occurred with a probability of 50 per cent, but unless they knew the secret of its state structure they would have no basis on which to unravel the 'system' it had adopted. In fact, if the robot were human we would have to say that it was cheating 'when it felt like it', and conclude that this state structure describes a robot whose behaviour is distressingly 'unmechanical' if a machine is defined as a device which behaves in an entirely predictable fashion.

This apparently arbitrary behaviour of the robot is, of course, very easily provided for by the inclusion of two 'if . . . then . . .' instructions in its program: 'If in State 1 and no input then do nothing and go to State 4' and 'If in State 4 and no input then do nothing and go to State 1' and one could well react to this illustration of mechanical 'free will' by saying 'so what?'. After all, home computers incorporate a program of this kind in order to generate random numbers, and there is a good deal more to human free will than an ability to toss a coin. Moreover, it is the programmer who has decided how the robot will behave – he may not in practice be able to forecast what it will do each time, but he has decided the probability that it will do one thing rather than another.

However, in terms of automata theory, 'knowing' something, such as what sequence of actions to take in response to a particular input, boils down to having the requisite state structure. In the cases that we have examined, the robots know what to do because they have been equipped with programs that tell them; their state structures have been designed by a programmer. In order to 'learn' what to do, an automaton would clearly have to be able to design new state structures for itself, or, in programming terms, to write its own programs. The question of whether automata can, in principle, perform this feat is of great importance to robotics. If robots will never be able to learn, then their usefulness will be limited by the ingenuity of programmers and any hope that they can teach us anything about ourselves must be abandoned, for, whether we are automata or not, it seems indisputable that we have the capacity to learn things.

But, this last assertion, that human beings have the capacity to learn new state structures, is not universally accepted at a fundamental level; indeed, the issue of its validity lies at the heart of a philosophical controversy which has been raging for the past decade or so. What is in question is the status of that feat of learning which is most generally held to distinguish humankind from the rest of the animal kingdom, our mastery of language.

The debate centres on the issue of whether, as language users, human beings are 'designed' automata, born with a state structure that facilitates the understanding of language, or 'learning' automata who acquire an understanding of language.

The first view is adopted by supporters of the American mathematician Noam Chomsky whose work first raised the issue. Briefly, Chomsky's philological researches revealed that in all languages it is possible to discern 'deep' grammatical structures which are very similar. Chomsky argued from this evidence that a state structure must be engrained in all language users which enables them to 'generate' statements which follow the grammatical rules governing the deep structure of language. If this state structure exists it is common to everyone, as well as being highly complex, and it follows, in Chomsky's view and that of his disciples, that it is inborn rather than learned.

If Chomsky is right the implications for robots are rather serious, for it would suggest that if we wish robots to understand any kind of language we will have to design them with very elaborate state structures from the start, and that their language using ability could only be extended by enlarging their programs and hence their state structures.

Chomsky's view is, however, hotly disputed by another school of thought developed by the Swiss psychologist, the late Jean Piaget. Piaget argued that language is a skill which children acquire as they grow up, just as they learn to use building bricks or to feed themselves with a spoon. Through meticulous observation Piaget was able to trace the stages in a child's development which marked the growth of linguistic skills and to show that these paralleled the phases in which other kinds of skill, which few would suggest are inborn, were acquired. If Piaget was right, it follows that the language learning robot may, in principle, be an achievable objective.

The Chomsky-Piaget debate is an example of an essentially philosophical problem to which robotics may, one day, be able to make a serious contribution; even though in investigating the origins of language it might seem to be straying a long way from its home base in engineering. Already, as we shall see in Chapter Nine, efforts are underway to program computers in such a way that they will be able to both use and understand natural language; thus far, it must be admitted, the results are at best of debatable value, and it may well be that the programming approach will give way to one based on unprogrammed automata. But the attempt to discover how best a robot can be given the gift of language may well, in due course, resolve the argument in favour of one or other party with a certainty that cannot be achieved by other means. With both sides deeply dug in, and ample supplies of ammunition to hand, the debate could continue to rage, unresolved, for years to come unless new evidence can be found. The outcome of a meeting between the two principals, backed up by supporters on both sides, which took place in France during 1976 would suggest that on the abstract level the issue cannot be decided one way or the other, for after several days spent in detailed argument the parties separated, each as firmly convinced as ever that the truth lay with their camp.

Coming back to rather simpler learning problems, it might seem that if the problem of devising a learning automaton is one of designing state structures which have the ability to design further state structures, or, in computer terms, devising programs which produce further programs, one possible approach would be to think of the process in terms of two automata, one of which designed state structures for the other. In practical terms this is quite conceivable since, as was pointed out in Chapter Four, the idea of 'nesting' one program inside another is a basic principle in computer programming. The difficulty is, however, that this solution only moves the real problem one stage further off: for how will the new automaton 'learn' a state structure that enables it to teach the original automaton? This question can only be resolved by introducing a third automaton that teaches the second automaton, and so on in an infinite regression. Chapters Eight and Nine, which deal with the progress that has been made in the field of artificial intelligence, will look at the success that has been achieved by those who have explored the possibilities of programming computers in this way. For the moment it is sufficient to note that, from the viewpoint of automata theory, no such hierarchy of automata, or programs, however extended, can dispense with an aboriginal 'designed' program or state structure.

Happily, there is a way of enabling an automaton to learn from experience which avoids the worst implications of this infinite regression, and we can begin to illustrate it by using the example of the bolt-sorting robot. It is important to qualify this immediately in order to avoid misunderstandings. First, far and away the most difficult task in terms of practical robotics would be that of teaching the sensing automaton to distinguish between satisfactory and substandard bolts, and it has taken engineers some fifteen years to develop a machine that can do this sort of thing by visual inspection. We will be looking at such a device in Chapter Ten. Similarly, teaching the robot how to pick things up and where to put them down is a separate issue; in practice it would be the business of subsidiary automatons, or subroutines within the program, to control the actions of arm and grippers. But it follows that if the robot can 'learn' the central process of remembering what kind of bolt it is holding and what it should do with it, then it can 'learn' the subsidiary processes as well.

The kind of learning we are talking about is very elementary, the sort of learning a child masters when he or she is taught that CAT spells 'cat' and MAT spells 'mat', not the much more sophisticated business of learning the difference between cats and other animals, let alone the kind of complexity involved in finding something out for oneself. We assume, therefore, that inputs are 'shown' to the robot as a teacher might write CAT on the blackboard, and that outputs are 'demonstrated' as a teacher would say 'that spells cat'. A little reflection will, however, show that the distinctions between this sort of learning and more advanced kinds are a matter of degree and that if an automaton can be taught the simple process we are interested in it should, in theory at least, be possible to build on that foundation. In fact, as we shall see, even a very simple learning process can lead to learning of another kind that is not immediately obvious.

The state diagram for the 'educated' or preprogrammed robot ready to perform the sorting process straightforwardly and without 'cheating' was this:

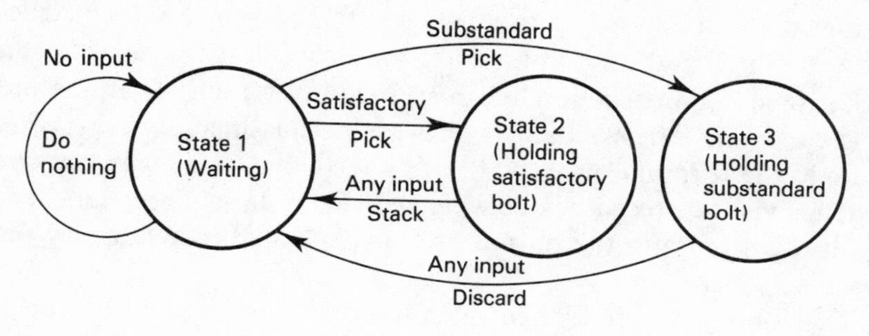

State diagram for the preprogrammed robot

The uneducated, unprogrammed robot has but one state, an inactive, neutral one, say State 0, in which it recognizes no inputs and produces no outputs:

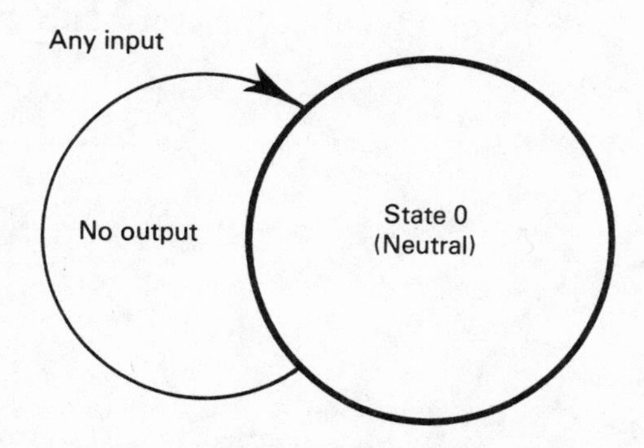

State diagram for the unprogrammed robot

In order to transform it into a learning automaton we have to equip it with a mechanism, a subsidiary automaton if you like, which performs a single function. It labels the states in which the main automaton finds itself (and also recognizes when two states are identical and combines them under a single heading). Technically, this secondary automaton is known as an 'X = Y' automaton since its purpose is to find a new Y (label) for each X (state) that occurs. We assume that it has available a series of labels in the form of 'State 1', 'State 2', 'State 3', etc and that it can be switched on and off by the 'teacher' as required. The 'X' which the mechanism is labelling with a 'Y' is defined by the combination of input and output which brings it about. Each 'Y', or label, is in fact attached to an instruction compiled by the automaton itself when it fills in the blanks of an 'if . . . then . . .' instruction like this: 'If ? (input) then ? (output) and X (new state).' The mechanism does not need to include the starting state in the formula since, by definition, the new state can only be reached via the preceding one which will have already been labelled. Both the input and the output are, as explained, provided by the teacher.

The robot starts in its neutral state, and if the 'teach' button is then pressed the X = Y automaton will swing into action and attach its first label to that state in which no input is present and no output is being demonstrated. In effect it is labelling the instruction 'If no input then no output and State 1'. If the teach button is pressed again in the same circumstances, the X = Y automaton will recognize them and conclude that the two states are identical. This is equivalent to formulating the instruction 'If no input then do nothing and remain in State 1', or of constructing this state structure:

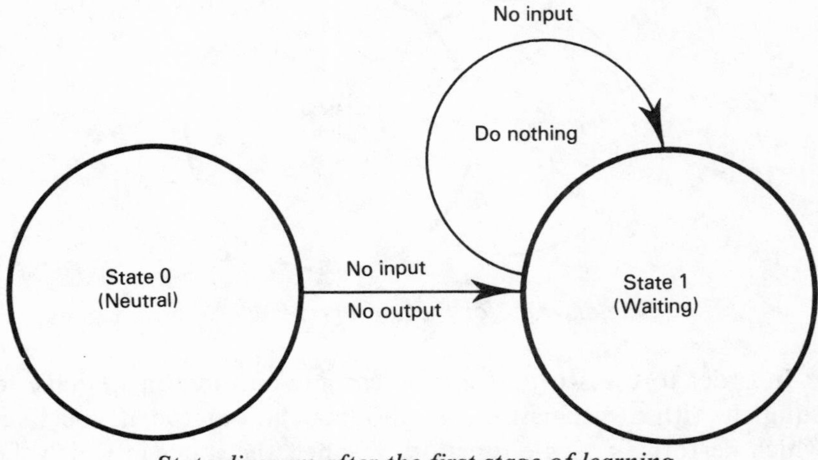

State diagram after the first stage of learning

If the robot is now shown a 'satisfactory' input and the task of picking up a bolt is demonstrated it will be in a new state, which the X = Y automaton can label as State 2, giving this state structure:

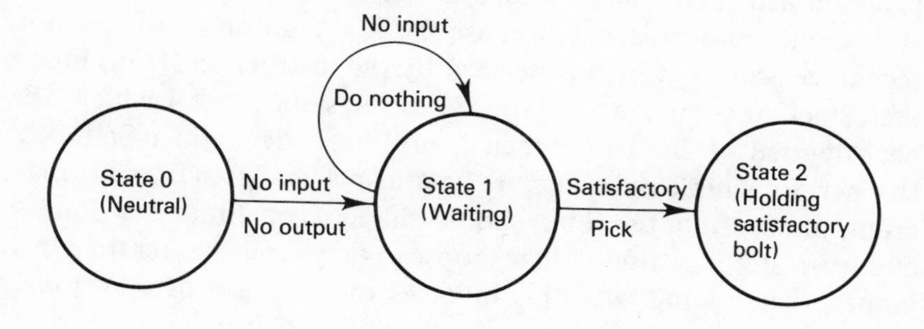

State diagram after the second stage of learning

The robot can now be 'taught', while being shown (no input) that the component should be stacked and the X = Y automaton will label the new state as State 3; finally, the automaton will be 'taught' while being shown no input and without any output being demonstrated. The X = Y automaton will recognize this state as State 1 and the first part of the state diagram will be complete.

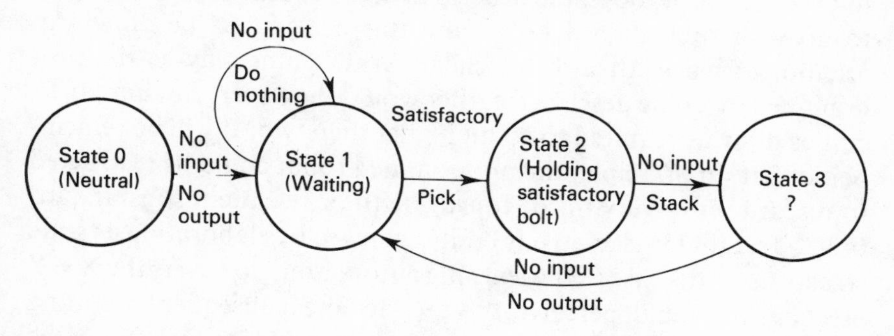

State diagram after the third stage of learning

Two points should be made at this stage. First, the 'if . . . then . . .' instructions that are being compiled are not formulated in words; they are simply sequences of logical operations in the robot's computer which relate inputs, outputs and states; and what the X = Y automaton (which is itself a part of the computer) is doing is giving those instructions addresses in the computer memory. The second point will already have been spotted by the attentive reader: the learning robot has a state structure for performing the 'satisfactory' part of its task which differs from that of the pre-programmed robot. There is an additional state, State 3. The extra state comes into existence because the X = Y automaton does not recognize State 3, which is defined by the instruction 'If no input then stack and change to State X', as the same State 1 which has been defined by 'If no input then do nothing and remain in State X'. (In fact, for simplicity's sake, we have also ignored, in the state diagram, the fact that the robot should change from State 2 to State 3 and from State 3 to State 1 in response to 'any input' rather than 'no input' since coping with this involves the creation of additional states.) It is only with a programmer, who knows in advance what state structure the automaton will require, that the short cut from State 2 to State 1 can be achieved. This is an instance of a general point: a learning automaton's structure will always include some 'redundant' states which would be unnecessary had it simply been programmed. Programmers, habitually in search of the most economic and elegant algorithm, find this worrisome, but it does seem that it is a condition of possessing even simple learning ability.

Returning to our teaching efforts, it would seem that all we have to do is to repeat the whole process in order to teach the robot what to do with substandard bolts, and this can indeed be done. An attempt to teach a robot to 'cheat' in the same way as the programmed machine described earlier would, however, run into difficulties if it was required to count higher than two. It will be remembered that, in principle, the programmed robot could be instructed to discard only every third, fourth, fifth or twentieth substandard bolt, a feat that was relatively easily achieved by elaborating its state structure. In the case of the learning automaton, however, the X = Y mechanism would perceive the second and subsequent cheating states as being exactly the same as the first one, for they would be defined by identical 'if . . . then . . . ' instructions. This snag can only be overcome if the learning automaton is first taught, as a separate operation, how to count. It will then have in effect a subsidiary 'counting automaton' which it can call upon as and when required. The necessity for teaching counting as a distinct, separate skill tends to confirm the evidence which child psychologists have accumulated to suggest that counting is a very fundamental skill in children, the elements of which must be mastered before a child can grasp any concept of plurality – that is, of several things being different from one thing.

But at this point an interesting question arises. If it were a human being who was being taught, we would not have to repeat the whole teaching process, for once he had been taught how to pick up a satisfactory bolt he could simply be told to do the same thing with substandard ones, and then shown how to discard them rather than putting them in a box. Supposing the human pupil to be distressingly simple-minded, but to have a retentive memory, obeying this order might require him to recapitulate mentally the steps he went through in order to pick up a bolt the first time. It is an interesting question whether the learning automaton could also recapitulate, internally, the conditions for the transition from State 1 to State 2 that are involved; for if it had this ability to 'rehearse' something in its head we would be forced to admit it was going through some kind of thinking process.

We shall not describe in detail how a robot would actually be taught in this way in practice, for it would involve going into technicalities that are difficult to follow, though easy to accomplish, and it is in any case a subsidiary consideration compared with the question of whether an automaton can be said to recapitulate, 'rehearse things in its head', or 'think'. It is, therefore, on this second point that we now concentrate. The key to the question is that any process of recapitulation involves having a picture of the world: of what is done to you as well as what you do, and of the relationship between the two. To put it in a slightly different way, recapitulation involves the acquisition of 'experience', the knowledge that particular actions have foreseeable consequences – knowing that if you do one particular thing the world will react in a certain way. It is this kind of experience which a child acquires when he realizes that knocking over a mug results in spilled milk – and probably an angry parent. Gaining experience, or learning about the world, is a process central to our growing up. The question is: has our robot learned anything about its world as a result of being taught how to pick up a satisfactory bolt and what to do with it?

The world the bolt-sorting robot has so far learned about is a very small one, consisting of the two possible inputs, 'nothing' and 'satisfactory', and it is a moot point, philosophically speaking, whether the world of a human being, which is of course much larger, consists of a finite or an infinite variety of such 'inputs' or their equivalents. However, even from this meagre material, the robot can be enabled to construct a world picture. The trick is to give it a mechanism which builds up, by a similar process of constructing and labelling 'if . . . then . . . 'formulas (which will now be 'statements' rather than 'instructions'), a picture of the world as a second automaton, the 'world automaton'. This mechanism, in other words, enables the robot to reconstruct from its own experience the 'program' which the world is following.

Go back now to what the robot was taught when it was shown how it should deal with satisfactory bolts. This can be set out as a sequence of inputs and outputs, starting from the very beginning when the robot was in the neutral State 0:

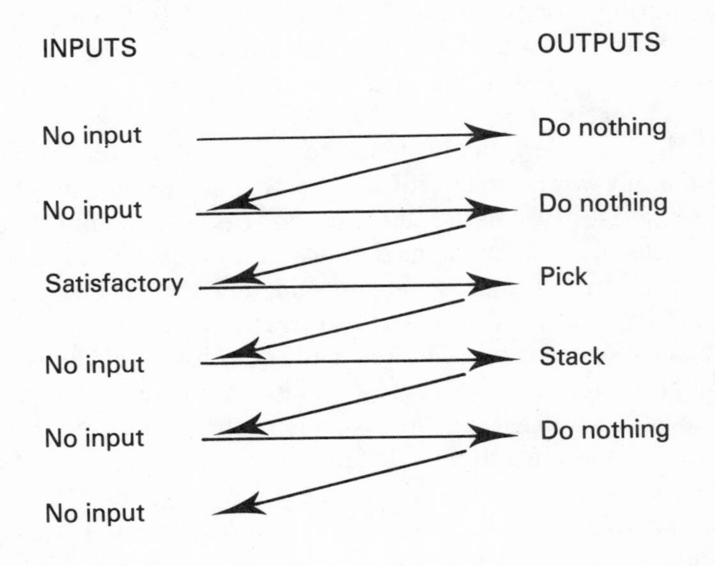

The robot's experience summarized

Using its logical facility for constructing 'if . . . then . . .' formulas, the robot can translate this sequence into a program which, from its point of view, defines the world automaton. The actions or outputs of the robot, of course, represent the inputs to the world automaton, and inputs to the robot represent the outputs of the world automaton. In practice it is simpler to think of the world automaton's outputs as both outputs *and* states. (The appearance of a satisfactory bolt, for example, is equivalent to both an output of the world automaton and a change in its state.) Thus the list of statements which defines the world automaton is like this (where State [and output] A of the world automaton means that the robot automaton is receiving no input and State [and output] B of the world automaton means that the world automaton is giving a satisfactory input to the robot):

If State A and no action by robot then State A

OR

If State A and no action by robot then State B

If State B and robot picks up bolt then State A

If State A and robot stacks bolt in box then State A

These formulas are, of course, 'if . . . then . . . ' statements rather than instructions, for no instructions inside the robot can *tell* the world what to do, but they can describe what it has done in the past and may, therefore, be assumed to be going to do in the future. If we translate these statements into a state structure for the world automaton the result will look like this:

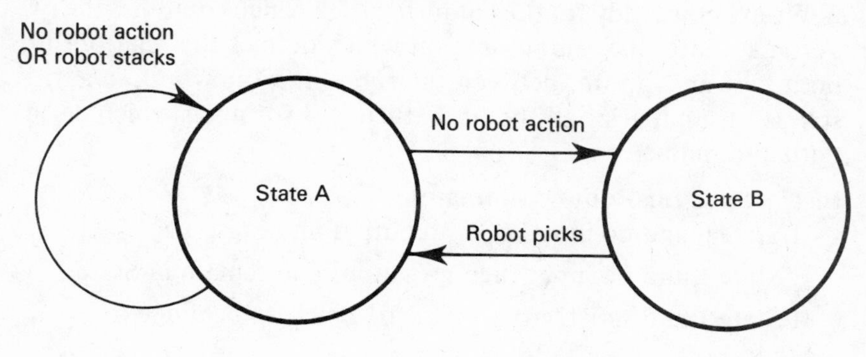

State diagram for the world automaton

The most striking feature of both the statements and the state diagram is that it is impossible to predict whether, when the robot does nothing, the world will stay as it is or present the robot with a satisfactory bolt. This reflects the fact that in the world at large it is always possible that one thing, or nothing, will occur rather than another; or in the robot's miniature world, that it can never predict when a bolt is going to appear. But it does begin to be clear that the process of recapitulation can take the form of an interplay between the robot automaton and the world automaton, both of which exist as programs inside the robot's computer. Overall the relationship between the two can be depicted like this:

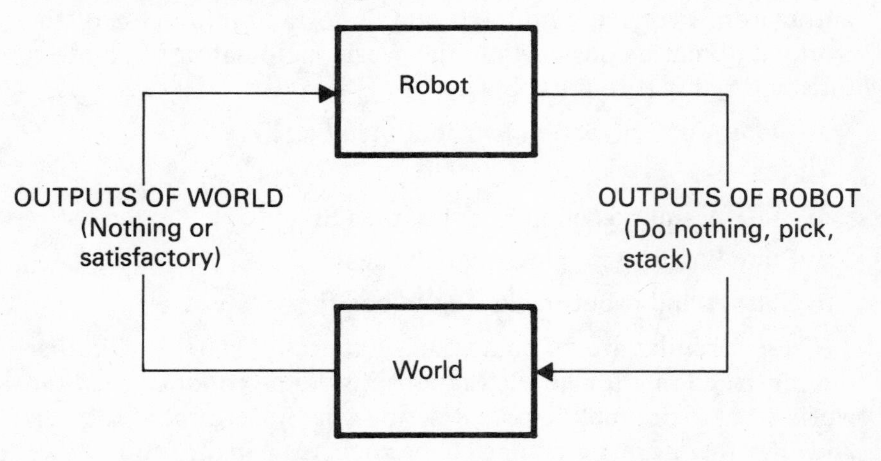

The relationship between the robot and the world

We are now ready for the conjuring trick which combines the two automata into one, an automaton which defines the relationship, both past and future, between the robot and the world. The first step is to summarize all the 'if . . . then . . .' formulas which define both automata:

Instructions for Robot Automaton

If State 0 and no input then no output and State 1

If State 1 and no input then no output and remain in State 1

If State 1 and satisfactory input then pick and change to State 2

If State 2 and any input put bolt in box and change to State 3

If State 3 and any input do nothing and change to State 1

Statements about World Automaton

If State A (no bolt present) and no input from robot then remain in State A *or* change to State B (bolt arrives)

If State B and robot picks up bolt then change to State A

If State A and robot stacks bolt then remain in State A

All possible states of the combined robot-plus-world automaton can now be defined in terms of three variables: the state of the robot automaton, the state (or output) of the world automaton, which is also the input to the robot automaton, and the output of the robot automaton, which is also the input of the world automaton. The combined automaton has therefore a state structure which defines not only all its possible state changes, but also all the sequences in which they can occur. Thus, starting from the point where the robot is in its neutral State 0, one of two things can follow once it has obeyed the 'if... then...' instruction which puts it into State 1: either the world automaton will continue to provide no input and the robot will continue to do nothing, or the world automaton will supply a satisfactory input and the robot automaton will pick up the bolt. These two eventualities can be shown by means of a state diagram for the combined automaton, and the process can be continued to cover all possible chains of events, giving a diagram like that shown overleaf.

Though at first sight the diagram is rather complicated, the fact that the robot's computer brain contains all information necessary to construct this state diagram for a combined automaton enables it to recapitulate 'in its head' every eventuality that can occur and the consequences of any robot action in any circumstances. The fact that in some cases the change from one state to another is not determined but subject to an 'OR' consideration can be dealt with by technical means which are relatively uninteresting – a little introspection will reveal that human beings readily devise means for coping with such indeterminancy, either systematically or arbitrarily. The point is that a state structure of this kind is different in nature from the sequence of instructions that make up a computer program, for it relies upon 'if... then...' *statements* about the world as well as 'if... then...' *instructions* for the robot itself. It allows the automaton to know what will happen to it as well as what it will do and can therefore be recognized as something very much like a 'thought' structure. Moreover, automata can traverse a state structure of this kind just as fluently and effortlessly as human beings think.

Looking at the state diagram it also becomes clear how, in principle, the state structure of the combined automaton can be enlarged to cover the contingency in which substandard bolts, which have to be discarded, can crop up instead of satisfactory

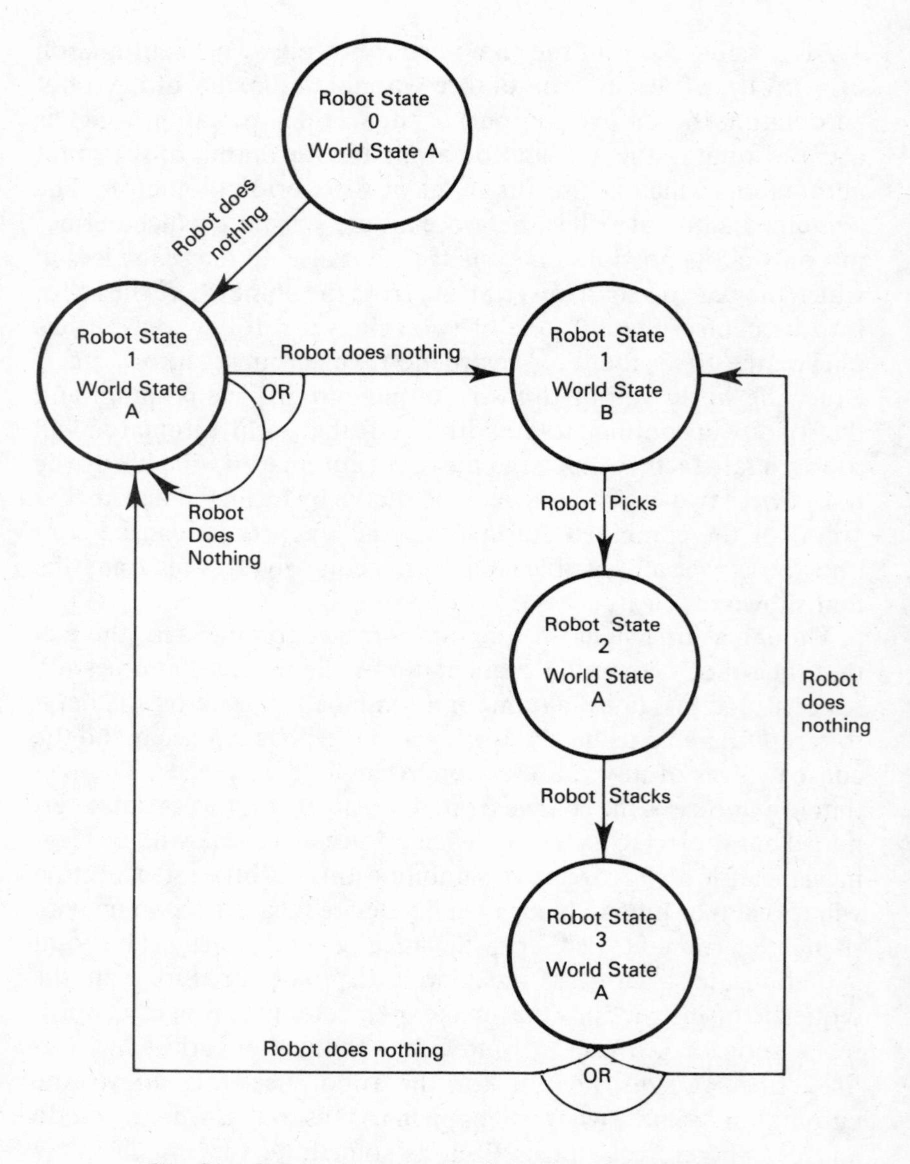

Combined state diagram for the robot and the world

ones, for it is evident at what points new 'OR' branches have to be added and what kinds of 'if . . . then . . . ' formulas will have to be provided for both the robot and the world automata in order that the combined state structure encompasses the wider range of possibilities.

To do a little recapitulation on our own account, it has been shown that an automaton with no 'given' or programmed state structure telling it how to behave can be taught one and that, in the process, it will learn a state structure for the world with which it interacts. This amounts to a demonstration that it is possible for automata to acquire experience, or knowledge of what the world will do to them, without any need for this to be supplied in the form of a program. The example used has, of course, necessarily been extremely primitive, but nonetheless it provides a credible model for the kind of learning process that a very young child goes through and, if we accept that it is possible to carry out the slightly more elaborate process of describing a new output in terms of a taught one, of the rather more advanced learning that results from being told things rather than simply being shown them.

It would certainly be possible to argue that, at the conclusion of the teaching process, the learning automaton we have described had a mind of some description. The memory circuits of its computer now contain information which was not there before, and which in all essential respects the machine has compiled for itself. Moreover, part of the information consists of 'experience' or expectation, and the machine is now driven to do the tasks it has learned by a 'need' which has been inculcated rather than preprogrammed.

Have we succeeded in finding the mind, or laying the ghost, in this particular machine? And if the catalogue of 'if . . . then . . .' formulas within an automaton's memory does constitute a mind of some kind, is it the prepackaged mind of the programmed robot or the self-constructed one of the learning robot which most nearly approximates our own minds?

At this point an important *caveat* has to be entered. The bolt-sorting robot whose mental processes we have explored in some detail is not a flight of fantasy, but a slight elaboration of the kind of 'playback robot' described in Chapter Five, of which many thousand examples are currently in use. But before leaping to the conclusion that our factories are full of machines with even very simple minds, we must remind ourselves of the kinds of brains which contain these minds, if minds they be. They are of course digital computers or microprocessors and in order for them to become learning automata they must be equipped with programs which tell them how to be learning automata. They can learn only because they contain programs which lay down the rules for compiling the 'if . . . then . . . ' formulas which define both the robot automata and the world automata and also control the operations of the X = Y mechanism that labels and files the 'if . . . then . . . ' formulas in the memory circuits. The methods used by the learning automaton may have cut short the infinite regress of automata, but only, it seems, at the cost of accepting the inevitability of some finite number of aboriginal programs.

It might be argued that if our robot has gained a mind of some kind, it is closer to Chomsky's idea of the mind than Piaget's, for although it constructed its own state structure, it did so under the tutelage of a programmed automaton.

This is a philosophical debate which, happily, we can avoid, for as will be explained in Chapters Ten and Eleven it has become possible to construct quite a different kind of information machine, one in which the ability to learn arises inevitably from the structure of the hardware involved. This kind of machine, once in existence, acquires a 'ghost', which can be analysed in terms of automata theory, as an inevitable consequence of its construction. In the final chapter we will examine the tantalizing possibility that this is, indeed, the kind of machine which human beings are.

8. Pre-packaged Intelligence

Here's a good rule of thumb:
Too clever is dumb.

Ogden Nash, 'Reflection on Ingenuity'

Whether or not it is dignified with the title of a 'mind', the automaton's faculty of determining an output by reference to a store of 'if . . . then . . . ' instructions clearly amounts to intelligence of some kind. The level of 'thought' manifested may be very low and, if it is entirely the result of preprogramming, may amount to no more than an extremely transparent stimulus-response mechanism; but nonetheless, it is difficult to dispute the conclusion that if a human being, or even an animal, behaved in the same way we would describe the behaviour as intelligent.

It would seem to follow that, if we grant that an automaton with a state structure which allows it to choose between different courses of action is intelligent in one way or another, then a learning automaton, with its capacity for being taught, is more intelligent than an entirely preprogrammed one. This argument is strengthened by the principle of constructing a world automaton, which allows the machine to know what is going to happen to it as well as what it is going to do, and seems to accord well with our own observations of how human infants extend their knowledge and understanding of the world by a process of widening their experience. But although roboticists have found the kind of learning mechanisms described in the last chapter to be an extremely economical means of teaching robots to perform simple, repetitive manipulations, where it is much quicker to 'show' the machine what is required of it than to try and program it on paper, the unprogrammed automaton has not found any favour, until very recently, among the computer scientists working in the field of 'artificial intelligence'.

In fact, the phrase, which was coined by John McCarthy of Stanford University in 1958, has taken on an entirely specific meaning and has become associated exclusively with preprogramming techniques. In retrospect, this is easily understood. In the 1950s the digital computer was just beginning to reveal its awe-inspiring potential. Its success in the fields of mathematics and logic was entirely due to preprogramming and it seemed self-evident that, if it was eventually to master other aspects of intelligence, the most profitable approach would be that of the innovative programmer. Wiener's suggestion that alternatives to the von Neumann computer should be explored was totally in eclipse at the time for, given the speed at which the von Neumann machine had mastered the kinds of intelligence for which formal algorithms already existed, it was reasonable to suppose that all other intelligent processes would succumb to its powers once they had been reduced to 'computable' form by the discovery of appropriate algorithms. Artificial intelligence (or AI) therefore came to mean the process of constructing programs for digital computers, or providing them with state structures, which enabled them to master, or mimic, intelligent behaviour.

The question of whether it is mastery or mimicry which is involved is central to the whole status of the field, and turns upon the issue of what is really happening when a computer is programmed to behave intelligently. The essence of the problem can be illustrated if we look at other applications of the computer's power. Suppose, for example, that an airliner crashes in circumstances which can be firmly established when its black box or flight recorder is recovered. It might well be decided that the easiest way of resolving exactly what happened and why would be to incorporate all the available information about the crash into a computer program which would allow investigators to 'reconstruct' the disaster. To be useful the program will not only have to incorporate all the data (wind speeds, the speed and altitude of the plane, the position of the controls, etc) but must also employ the right 'model', that is the relationship between aerodynamic cause and effect must be accurately reproduced.

Now when that program is run what takes place inside the computer is an event related to the original crash but of a quite different order. The crash itself is an event of the first order, a happening in the physical world; inside the computer there occurs an event of the second order, a model of the first.

The question obviously arises as to whether, when we witness a computer running an artificial intelligence program, we are watching an event of the first order or an event of the second order. The issues this raises are fundamental, for first order events are, by definition, things that happen in the present and can only be described once they have happened; we cannot know what we are going to think until we have thought it – unless, of course, our own thinking is an event of the second order rather than the first. Events of the second order are, however, determined in advance; given complete and accurate information the computer can arrive at only one conclusion about the air crash. Thus, even if a computer 'proved' its intelligence by passing a Turing test (see Chapter One) one would still be entitled to ask whether what had occurred was a real event or a model of reality; rather as we might ask, if we saw an airplane crash on television, whether we were watching a real airplane crash or a spectacular special effect contrived by use of a model airplane.

There is an argument, which was forcefully advanced by Turing himself, that in the case of thinking the distinction between events of different orders is meaningless; all that matters is that the appearance is the same. If a computer seems intelligent then to all intents and purposes it *is* intelligent. Even if, for the moment, we accept this argument, two further questions arise. First, is a digital computer on the von Neumann model a machine which, by its very nature, can reproduce the whole range of behaviour which we categorize as intelligent? We defer discussion of this issue until the next chapter, simply because it becomes most significant in examining the kinds of intelligence required to interpret visual information and understand language, and turn instead to the second question. Are the models of intelligence which AI has adopted valid ones?

This issue is obviously crucial to the whole endeavour, for just as the crash investigators would have learned nothing about the real crash if the model they used was inadequate, so artificial intelligence research will tell us nothing about 'real' intelligence unless the model it uses is appropriate. It would naturally be unrealistic to expect that programmers should have set about producing a set of instructions which encompassed intelligence in all its variety, and scientists concerned with AI have in fact adopted a number of separate approaches, each of which attempts to devise a model of a particular aspect of intelligence. One of the most interesting of these, and the one with the longest history, is what has become known as 'game playing'.

The ability to play chess at a reasonable level was one of the most popular targets which the early computer pioneers set for their primitive machines and it is relatively easy to see why this was so. Chess is a game that indisputably requires intelligence on the part of the player, but its rules, the world within which the intelligence must be applied, are clear and well-defined; it is a 'game of perfect information' (that is, there is no element of luck or bluff as in poker); and the information about the game is easily codified into digital form – even human chess players are accustomed to shorthand of the 'P-QN3 N-B6' kind. Chess, or other board games, can also take advantage of the computer's basic ability to deal with 'if... then...' questions. For one of the fundamental skills in chess is the ability to calculate the implications of a particular move for the future and such calculations are obviously of an 'if... then...' nature. The actual mechanism employed takes the form of a 'decision tree', where each main branch represents one of the moves open to a player and then subdivides into a number of further limbs to show the answering moves that can be made by his opponent and so on. Naturally, the player who can look furthest ahead along the most branches will have an enormous advantage.

The complexities of chess produce a decision tree which rapidly spreads to unmanageable dimensions; but we can use a much simpler game, noughts and crosses or tic-tac-toe, to demonstrate the principles involved and some of the techniques which have been developed to deal with more complicated games. As is reasonably well known, noughts and crosses is a game in which, providing certain rules are followed, the player who has the first move can always ensure a draw. Briefly, the first player can always achieve a situation, after his second move, like those below where there are five axes on which a row of x's can be completed within two x-moves.

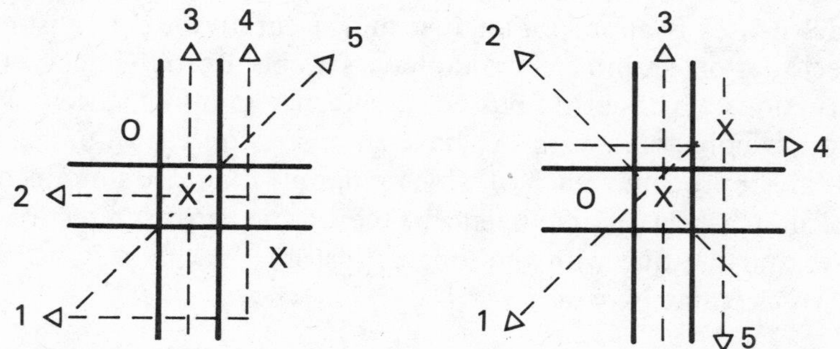

Positions after two moves by X, showing the five axes on which a row of X's can be completed within two moves

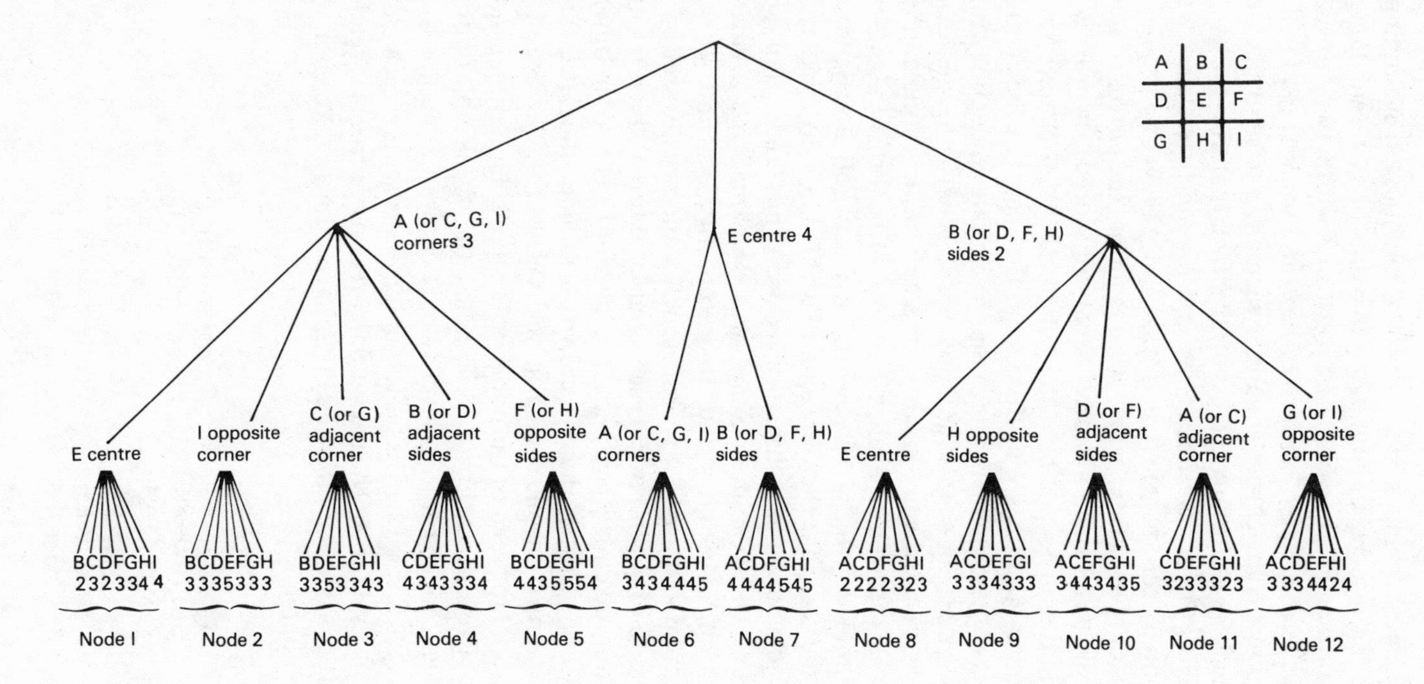

The noughts and crosses decision tree

This suggests a 'scoring' system which 'x' might use in evaluating his choices at each stage, by counting one point for each remaining open axis on which he can complete a row of x's within two moves. On this basis, as the reader can verify, the centre square on a blank grid would score 4, the corner squares 3 each and the central, side squares 2 each. Taking advantage of the fact that for the two opening moves, the orientation of the grid remains irrelevant, we could draw up a decision tree (see previous page) covering the first three moves, or levels of the game. Note that only the moves of the starting player (level one and level three) are scored. In order to identify them the nine squares of the grid are denoted by letters.

As it happens, in the case of this game, the fact that the centre square scores highest on the first move is an accurate guide, but with more complicated games this may not be so. What is needed is a method for assessing the implications of a move at level one on the range of choice that will be available at level three (or, if the tree is elaborated, at level five or level seven). This must, obviously, take account of the fact that the opposing player is likely to make the move most advantageous to him at level two. The basic technique employed, known as 'mini-maxing', was developed for a chess playing program in 1950 by that ubiquitous figure in this story, Claude Shannon. To apply it to our noughts and crosses tree, we take each group of seven possible scores at level three which result from a decision taken by the opposing player at level two, in other words, the group of scores that are shown on a single 'node' (the fan of lines emanating from one choice on the level above). There are 12 such groups, one for each of the nodes of the tree at level three, and the *highest* figure in each group is called the 'back-up' figure because, when 'backed up' to level two, it shows the choice the opposing player faces if he wishes to minimize the score which can be achieved at level three. The back-up figures on our tree are, reading from left to right and grouping them according to the choices made at level one:

Node 1:4
Node 2:5
Node 3:5
Node 4:4
Node 5:5
} Corner square chosen at level 1

Node 6:5
Node 7:5
} Centre square chosen at level 1

Node 8:3
Node 9:4
Node 10:5
Node 11:3
Node 12:4
} Side square chosen at level 1

The inference can now easily be drawn that, if the first player places his mark in the centre square at the outset, there is no way in which his opponent can prevent him from obtaining a maximum score of 5 at level three; if, on the other hand, the opening move places a mark in a corner or side square, then the opponent can always prevent a maximum score at level three by making the right choice at level two.

Mini-maxing ignores all the scores at intermediate levels. Starting with the scores at the final level, it backs them all the way up the tree, taking the maximum score at the levels where, it is assumed, the program makes the highest scoring choice, and the minimum scores at the intervening levels where, it is assumed, an opponent will make the choice least advantageous to the program. The diagram overleaf shows how mini-maxing would work on a very simplified tree working up from the fifth level. The heavy line shows the route through the tree that a mini-max program would follow.

Unfortunately, the simplest of games produces a decision tree a great deal larger than this – even a noughts and crosses or tic-tac-toe tree, based on the three levels which are explored above, would sprout over 60,000 twigs (each of the 72 branches at level three sprouts six more branches at level four which, in turn, each subdivide into five more at level five, etc)! If a game which most six year olds can master with a little practice gives rise to so large a range of choices, it can be imagined how rapidly a chess tree spreads to unmanageable proportions. Even for a computer, running through thousands of possibilities every second, the task of assessing a chess game several moves ahead can quickly become time consuming. Readers who have played against machines which can be adjusted to play at different levels of skill will have noticed that the higher the standard demanded of the machine, the longer it takes to decide on a move; this is a reflection of the fact that it is searching decision trees which reach further ahead and that with each increase in the

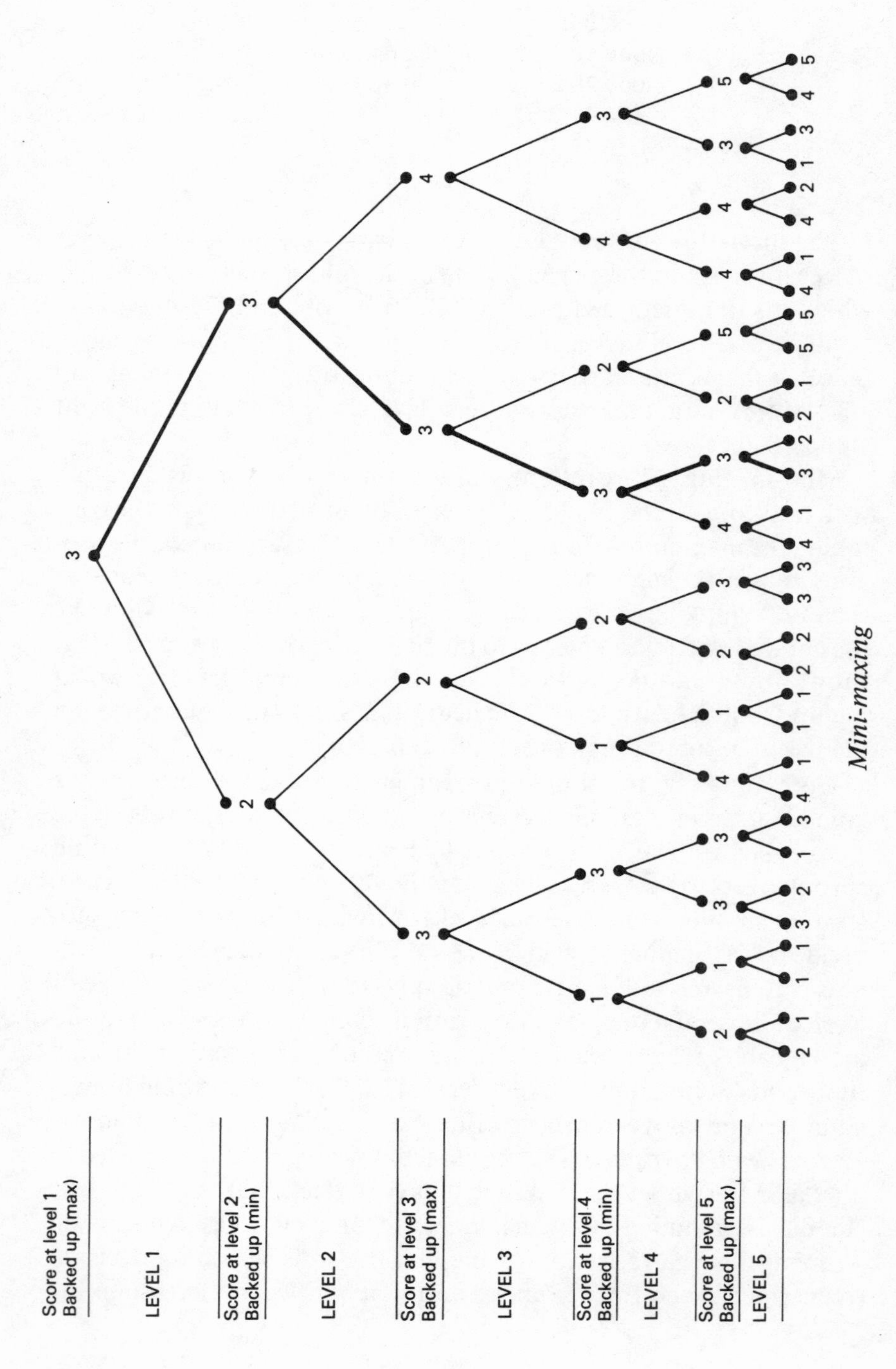
Score at level 1
Backed up (max)
LEVEL 1
Score at level 2
Backed up (min)
LEVEL 2
Score at level 3
Backed up (max)
LEVEL 3
Score at level 4
Backed up (min)
LEVEL 4
Score at level 5
Backed up (max)
LEVEL 5

Mini-maxing

number of levels searched the range of implication to be examined becomes larger by several orders of magnitude.

Mini-maxing quickly became established as the basic game-playing technique, and most subsequent work has concentrated on devising ways of simplifying the decision trees and cutting short the process of assessing and backing up the scores. Shannon himself invented a process called 'alpha-beta-ing' which goes some way towards limiting the work involved in backing up figures. Although the mathematical rules are a little technical, the principle involved is clear: the search beneath a particular node is abandoned when it becomes clear that the 'maxes' and 'mins' it is going to produce are less favourable than those found beneath nodes which have already been examined. One obvious drawback of the alpha-beta process is that its effectiveness is largely a matter of chance – if, for example, the search moves from left to right along the scores at the final level, then if the optimum choice happens to lie in the leftmost node, searches on other nodes will be abandoned relatively quickly, but if the rightmost node contains the optimum choice the whole tree may have to be searched in order to find it.

Chess playing programs were further refined as a result of the work of A L Samuel of IBM in 1959. The first improvement Samuel brought about was a result of limiting the scope of the decision tree originating with any particular move. A human player, even at tournament level, cannot and does not attempt to calculate the detailed implications of moving a pawn on the extreme left hand side of the board for the play on the extreme right hand side, for, clearly, these will not become evident unless one looks, perhaps, ten or a dozen moves ahead. Samuel developed rules for testing the consequences of a move in terms of the immediate neighbourhood of the piece to be moved, the surrounding eight or twenty-four squares on the board, say. This involves looking at the 'state' of the game in that particular area and that area only, and ignoring the state of the game as a whole.

The advantages are obvious, but the drawback is that 'thinking' in this way causes the program to concentrate on tactics at the expense of strategy. It is this handicap that has probably prevented chess playing programs from advancing to the stage where they can consistently succeed at grand master level. For although human players of world standard almost certainly cannot look ahead as far as a sophisticated computer program, they do seem to have an instinctive ability to remember and recognize winning patterns in the game as a whole and without going through an elaborate process of computation. In other words, they can recognize a 'winning state' on the board, as a whole, and the states that lead to it, without exploring the decision trees concerned. Nonetheless, it seems certain that programs will eventually be constructed which will be able regularly to defeat grand masters. But when this happens it will be because sheer increments of computing power have enabled the program to look ahead in both depth and range to an extent which the human brain cannot achieve. In some sense, therefore, the computer will be playing a game which is different from the game which humans play.

The second area which Samuel tackled was the all important one of scoring systems. In constructing the noughts and crosses program, we were able to take advantage of the fact that the game is a determined one and a scoring system for level three is relatively easy to devise. But in chess, or even in simpler games like draughts or checkers, this is clearly not so. By level three, or level thirty, a chess game may be far from decided and it is impossible to fit decision trees which cover the entire duration of a game into the confines of even the largest conceivable computer. It is, therefore, necessary to construct rules of thumb by which to evaluate, or score, the consequences of a particular move. In checkers, for instance, one might decide that a move forwards by one square was worth one 'point', that taking an opponent's piece was worth three 'points' and that losing a piece of one's own was worth minus three 'points'. In chess, as was seen above, the merits of a move are more likely to be evaluated in terms of the state of the board which results from it.

Arbitrary rules of this kind, which have to be employed in many branches of artificial intelligence, are known as 'heuristics'. In the context of AI, the word is used in a slightly different sense from its dictionary definition, which describes an heuristic as a piece of knowledge based on experience: for example, a 'dictionary' heuristic for driving from London to Edinburgh would be, 'follow the route you have travelled before,' whereas an AI heuristic might take the form, 'whenever you come to a junction take the road that runs most nearly due north.' Samuel's work aimed at making the heuristics involved in chess playing more like the dictionary kind by finding ways in which programs could learn from experience.

The benefits that would result are obvious. In a game as complex as chess it is virtually impossible to invent heuristics which hold good in every possible situation; losing a pawn, for example, may be a 'good' or a 'bad' thing depending on the circumstances. What is needed is a way in which a program can gain experience and learn to recognize the significance of a state of the board, or of a particular area of the board. The mechanics of scoring are relatively easily achieved in programming terms. Any position of the pieces can be reduced to a logical formula and the program can be taught to attach high scores to those positions which are 'winning' positions and low ones to 'losing' positions; or, taking the thing a stage further, to score highly positions which lead to winning positions and so forth. One interesting difficulty does, however, emerge concerning the teaching process itself. A program can only increase its knowledge by watching, or playing against, a human player (though knowledge gained by one program can, of course, be subsequently incorporated in replicas).

The problem arises because the only 'real' value of a move or a board position is that it leads to winning, or checkmate; but, unless the endgame has been reached, checkmate lies at the extremities of a decision tree which stretches out beyond the reach of the program. Thus the program can only learn by noting all the positions that occurred in the course of a game and then, when a result has been achieved, attaching a positive value to those which are revealed to have been winning ones and a negative value to the losing ones.

Unless the program is to move its pieces about the board by some random process, it must be equipped at the outset with at least some heuristic before it can begin to learn and if two programs are set to work playing against each other the process will generate no new knowledge in either. Both will be scoring their moves according to the heuristics which they have been given; neither knows what winning *means* because, unless the game fortuitously comes near to checkmate, winning is outside the range of their decision trees. Two such 'uneducated' programs will behave rather like the generals of the First World War, who became so preoccupied by their self-imposed heuristics which defined victory or defeat in terms of a few hundred yards of ground gained or lost that they often seemed to lose sight of their real objective, winning the war.

Harking back, briefly, to the description of an unprogrammed, learning automaton in the last chapter, it is important to realize that the kind of learning process which a chess playing program undergoes is far more trivial and superficial than that involved in learning an entire state structure for both robot and world automaton (page 149) which the learning automaton constructed for itself as it was taught. It should be clear that a state structure is 'given' to a chess playing computer in the form of its program, which, since it describes the rules of the game (which govern the behaviour of both the machine and its opponent) and the geography of the board, includes all the 'if . . . then . . . ' formulas for the machine and the world with which it interacts. But it will also be recollected that the state structure for the combined automaton contained instances where it might change to one state OR to another. The heuristics that a chess program learns are essentially rules for determining which branch the machine should favour when it faces such an OR question. To put it another way, a decision tree is simply an alternative form for a combined state diagram, and the addition of a learning ability enables the machine to refine the scoring system it applies to the diagram when faced with an OR.

In so far as the hopes of the early pioneers have been realized, indeed exceeded (for few in the early 1950s could have foreseen the day when it would be possible to buy a chess playing computer which could more than hold its own at tournament level for a fraction of the cost of a new car), game playing must be accounted a success for artificial intelligence. What remains to be shown, however, is that the model of intelligence on which game playing is based has any utility when it is applied to a world less constricted and codified than the chess board. We shall return to this question shortly, after looking at another, related, branch of work in AI.

However intriguing and challenging the possibilities of teaching computers to play games, it is clear that the 'thinking' required is a rather limited instance of a more general kind of intelligence – what might be described as the ability to solve problems. The possibility that computers might be capable of problem solving of a wider, less specific nature has great relevance to the future of robotics. If, for example, robots were to be employed in assembly work performing tasks now carried out by human labour, it would be extremely helpful if, rather than all components having to be delivered in consistent orientations, the robots had the intelligence to pick them up from a pile and align them correctly before putting them together.

Two typical examples show the kind of problems that might require solution. If a nut is lying on its face and the bolt it must screw on to is sticking out horizontally, which is the most economical way to manoeuvre the nut so that the two threads are correctly aligned? How should a shaft be manipulated in order to fit the end into a bearing?

It is because of its practical relevance to questions like this that most problem solving work in artificial intelligence has concentrated on what are known as 'robot task' problems. The kinds of problems tackled, however, generally fall far short, where complexity is concerned, of the nut-bolt, shaft-bearing ones mentioned above even though these might seem trivial to the average handyman. The 'objects' that most computers manipulate in AI laboratories are representations of simple, often one-dimensional shapes and the robot arms are themselves hypothetical. Perhaps the best-known program of this kind is SHRDLU, the invention of Terry Winograd, then at Massachusetts Institute of Technology, which enabled a computer to describe to a human 'interrogator' what it saw in a tiny world consisting of blocks and pyramids and how it would go about rearranging them to achieve some prescribed result.

Some of the problems tackled by programs like SHRDLU, and the way in which they are dealt with, can be illustrated by use of a very simple example. Suppose that a robot is 'shown' a table with a pile of three coloured blocks, stacked one on top of another in the order, from the top, red, white, black, and it is required to work out the most economical sequence of moves in order to restack them in the order red, black, white. For problem solving purposes, where the robot arm and its movements exist only in theory, all the computer

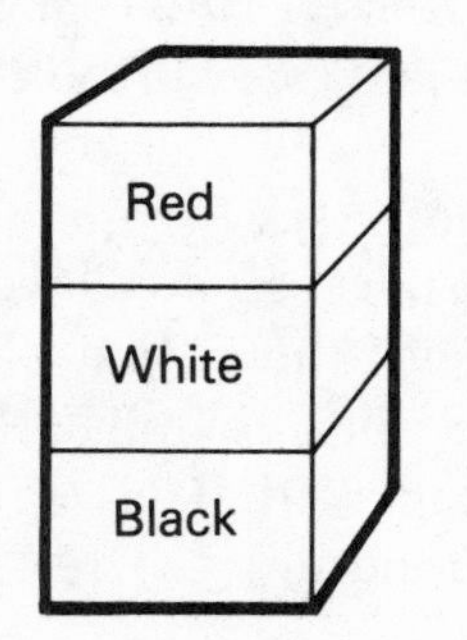

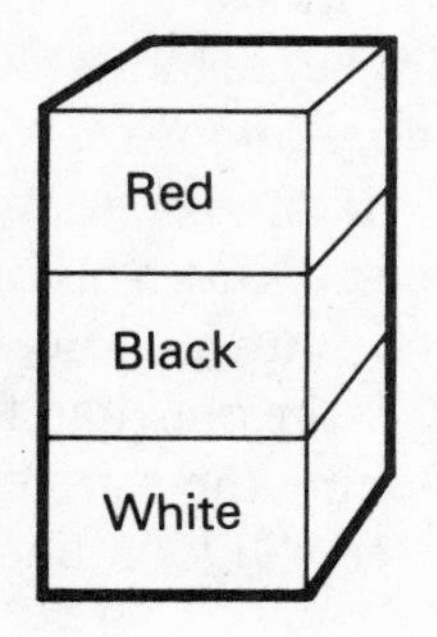

needs to know are three basic outputs: pick up a (named) block, put the block that has been picked up on top of another (named) block, or put it on the table. It will also be given some simple rules: it will be told, for instance, that only blocks which do not have other blocks on top of them can be picked up or have other blocks placed on top of them.

One way of dealing with a problem of this kind is by the use of decision trees, as in game playing. In this case, only one opening move is possible, but by level four seven possible actions have to be reckoned with:

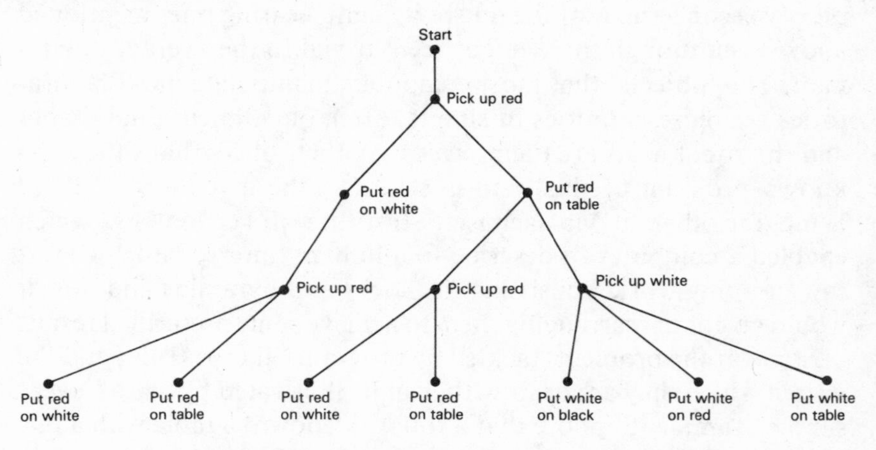

First four levels of the forward looking decision tree

There is, however, a difference between this problem and those that arise in game playing: for there is one, and only one, satisfactory conclusion to it. This factor makes it possible to build a decision tree that works backwards, starting from the situation which it is desired to reach. The first three levels of that tree will look like the top figure opposite.

Now if the program has some means of assessing these 'backward' moves in terms of their proximity to the goal, it will recognize that the fourth choice from the left at level three, 'put black on white', is the one that it needs and it can pursue this limb of the tree, abandoning the others. Thus, continuing to a fifth level, the backward tree will look like the lower figure opposite.

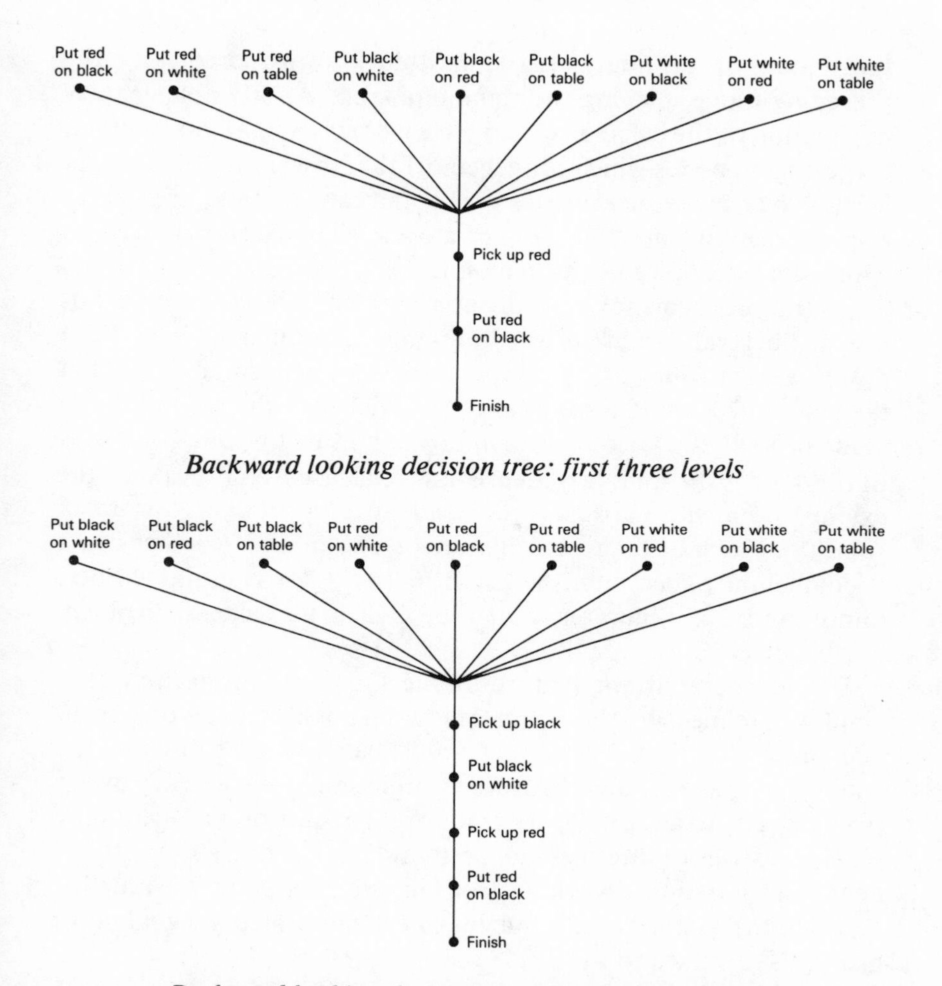

Backward looking decision tree: first three levels

Backward looking decision tree carried to level five

At this point, if the program refers back to level four of the forward looking tree, it will be able to recognize that the rightmost branches of the two trees have linked up and that the problem can be solved by the sequence: pick up red, put red on table, pick up white, put white on table, pick up black, put black on white, pick up red, put red on black. The sort of mechanisms required to solve problems in this way can also be formulated in terms of a specialized branch of logic called predicate calculus. In fact, we used a very simple example of the kind of considerations the calculus deals with in constructing the backward tree, for we made the assumption that the action of putting a particular block down predicates the preceding move, picking the same block up. Another elementary application of the calculus would have led our program to abandon the left hand nodes which branched off the forward looking tree at levels two and three because they lead, and can only lead, to alternation between two possible pairs of moves: pick up red – put red on white and pick up red – put red on table.

Several programs for working at a problem by the simultaneous use of backward- and forward-looking decision trees have been developed for robot use; perhaps the best known is that devised in the early 1970s by R E Fikes and Nils Nielsson which goes by the name of STRIPS. One of the main difficulties involved in programs of this kind is the finding of heuristics which accurately evaluate the extent by which a particular move advances a problem towards a solution. Much of the research in this area can be traced back to a fundamental paper published in 1960 by Alan Newell, Herbert Simon and J C Shaw on what they called the General Problem Solver, or GPS.

The most important feature of the GPS was the technique it employed to measure the distance between possible states of a problem and its solution. The idea can most easily be illustrated by the analogy of map reading. When contemplating a long journey by car from one city to another, a driver may plot a route on a road atlas by picking a series of intermediate destinations, towns, say, or prominent road junctions, working from his starting point towards his destination, so that the attainment of each intermediate goal diminishes the distance to the final goal.

The GPS used similar methods to identify intermediate goals which brought a problem closer to its solution by a measurable extent. There are classes of problems for which this method works well: if, for example, it is necessary to move a stack of ten identical bricks from the left hand side of a table to the right hand side, then each time a brick is moved the problem will be one tenth nearer a solution. There are also problems which will not yield to this approach; it would be futile, for instance, to plot a course through a maze by drawing an imaginary line from entrance to exit and marking sub-goals along it; similarly, one cannot measure success in reassembling a Chinese box puzzle by counting the number of bits that remain loose, for the order in which they are put together is all important.

One of the striking things about the history of efforts to devise workable AI programs, whether game-players or problem-solvers, is that most of the fundamental work was done very early on, in the 1950s and 60s, by pioneers such as McCarthy, Shannon, Samuel, Winograd, Nielsson, Donald Michie of Edinburgh University and Marvin Minsky of MIT and that subsequent advances have largely consisted of tidying up and refining the methods they originated. The most significant consequence of this has been that optimism about the potential of AI which was widely, and perhaps somewhat foolishly, expressed in the 1960s has been replaced by caution and even pessimism. The fact that the bullish attitudes which prevailed in the 1960s turned out to be premature is largely due to the fact that workers in the AI field underestimated the problems of 'scaling up' artificial intelligence techniques that had proved successful in the 'model' worlds made up of chess-men or coloured building bricks to cope with the problems that arose in the 'real' world. The point at which the bubble of optimism burst can be dated with some accuracy. It happened in 1973 and the man who applied the pin was a British mathematician, Sir James Lighthill.

Lighthill was appointed to produce a report for the Science Research Council, the body which funds most university research in Britain, on the prospects of contemporary artificial intelligence work. The weakness which he identified and homed in on with considerable vigour was the failure, in his view irremediable, to produce methods which had value in the real world. And the cause of the failure, and the problem which he argued was insurmountable, he christened the 'combinatorial explosion'.

The combinatorial explosion is a consequence of the fact that, although we are accustomed to think of things like chess problems or the more elaborate kinds of brain teaser as requiring more 'intelligence' than the business of everyday life, the reverse is really the case. While it is true, for instance, that the combinations of possible positions and moves on the chess-board are, for all practical purposes, infinite, the choices open at any one time are limited by the rules of the game. If we turn to a familiar household chore like dressmaking we find that while the 'combinatorial' problems are almost as great, the rules that can be used to resolve them are not nearly so well defined.

A dozen pieces of fabric forming a dress pattern, each with, say, four edges, can be stitched together in over 16 million different ways, and this is before taking account of the fact that there are hundreds of thousands of differing dress designs, each available in a whole range of sizes as well as variations in materials, colours and patterns. And the rules about what shape a sleeve should be or how a skirt is gathered at the waist are far more flexible and ill defined than those which govern the movements of chess-men. The world of the dressmaker, therefore, is not a tidy, self contained one like the chess-board, but includes these myriad and often subtle distinctions.

It would be relatively easy, of course, to program a robot dressmaker to produce a limited range of garments, operating by rote. But this could scarcely be described as an intelligent task – an intelligent robot would have to be able to take the same starting point as a human dressmaker, that is to say it would have to identify each piece of fabric by matching it to the pattern and be capable of following general instructions such as 'pleat' or 'hem' in a way that produced an acceptable result. All this would require that it have some understanding of the principles involved, of what a sleeve or a belt is, for example, and how it fits into the overall concept of a dress. Elementary though these ideas seem to most of us, they lie well beyond the reach of any artificial intelligence program devised to date. For just as the possibility of checkmate cannot be envisaged by a chess playing program until its accomplishment is a matter of a relatively few moves, so the general concept of a dress, as the common product of many different combinations of a whole variety of pieces of fabric, cannot be defined as the goal of any decision tree of remotely manageable proportions.

The implications of the combinatorial explosion are most disastrous for the problem solving model of intelligence, for the real world of the factory is usually infinitely more complex than the simple world of cubes and cones with which AI has concerned itself. The difficulties are compounded by the problems (discussed in the next chapter) of constructing programs which can actually recognize and identify visual images, such as those of the bits and pieces that go to make up a product. In engineering or electronics, say, components are not normally simple cubes or pyramids and do not routinely present themselves in neat ranks, all orientated in the same way – unless, that is, automation techniques have been used to ensure that this happens.

The combination of vision and intelligence would certainly be indispensable to a robot which was to master a skill that comes easily to any human worker, that of picking one particular part out of a random assortment. Yet this problem, the so-called bin-picking problem, is so severe that its solution has become a kind of *ultima Thule* of robotics, a tantalizing but elusive goal. The difficulty is not one of physical dexterity or manipulative skill, it is simply that it has proved difficult, so far, to find a model of intelligence that accounts for our ability to recognize a three-dimensional object, whatever its orientation, to distinguish it unfailingly from its surroundings and to calculate rapidly the best way of manoeuvring a hand to grasp it.

The consequences of the realization that the combinatorial explosion may limit the game playing or problem solving techniques of artificial intelligence to a world of games and trivial manipulations extend well beyond the utilitarian: for the fact that human dressmakers or bin pickers cope effortlessly with the kinds of difficulties which AI finds insuperable suggests that AI techniques are, at best, only a partial and limited model of the way in which our own intelligence works. It looks rather as if decision trees, predicate calculus, etc may only be rationalizations of information processing feats which are, in reality, achieved by quite different means – a sort of gloss which the brain produces in order to explain its own behaviour.

Why, it might then be asked, does the combinatorial explosion leave human beings so relatively unscathed? It seems that there are three separate, but often interlinked, factors involved. First, the fact that we learn slowly and gradually by experience, or in terms of automata theory, that we build up a world automaton step by step and, at the same time, shape and refine our own state structure to fit in with it, allows us to accumulate intelligence of a flexibility and subtlety that could never be acquired simply by 'book learning'. To see the force of this argument, it is only necessary to consider the improbability of a human being learning quite a simple skill, bricklaying, say, or plumbing, entirely out of a book without having the benefit of any practical experience at all – the chances that his first brick wall or central heating system would turn out to be wholly successful are patently rather slim. Yet this is exactly what we expect of a preprogrammed robot, and we make no provision for it to learn from its mistakes when it does try to put its knowledge into effect.

It seems more plausible to believe that the skills which a human being may only master after much trial and error or a long apprenticeship would only come easily to a machine which learned them by some comparable process – in other words, a skilled automaton is likely to be a learning automaton. Even in the case of the very trivial skills with which AI is currently concerned, the same sort of consideration might be expected to apply. Human infants, after all, do not leap from their cradles one morning and master a set of instructions for feeding themselves with a knife and fork or building model skyscrapers with their building bricks. On the contrary, they have to undergo a relatively long and painful learning process, during which their dinner is as likely to end up all over their faces as in their mouths and it is a red letter day when they achieve a pile of blocks more than three high.

The second advantage we enjoy, which is linked with the acquisition of experience, is our ability to generalize. We seem to enlarge and refine our own 'programs' for life, the sets of general rules about the way the world is, by a process of induction; that is to say that only after we have seen a number of specific examples or instances do we conclude that they illustrate some general point. The way in which we educate young children takes account of this. We do not show a child a single picture of a cat and expect him to remember for evermore that this is what cats look like; instead we assume that the notion of a cat will only become firmly fixed in the child's mind after he has seen several pictures of cats, or indeed several cats.

In the case of a programmed machine, however, we make exactly the contrary assumption. For the whole point of a program is that it aims to be a complete set of general rules to which the machine can refer whenever it encounters a particular instance. This is, of course, the only option open to us within the programmed model of intelligence, for a process of inductive learning would have to be based on experience. The problems which result from the use of exclusively deductive methods become particularly evident in the case of vision, which is examined in detail in the next chapter. For the combinatorial implications of trying to include in a single, workable program a set of rules which allow a machine to interpret every image it receives from even a comparatively restricted world are extremely grave.

The final factor is the absence, in programmed machines, of any intentionality, of any idea of *purpose*. The fact that, as we have seen, it is impossible to instil the notion of winning in a chess program is but one instance of this. It becomes an even more serious handicap in the case of language using programs, for the purpose of language is to convey meaning and though it has proved possible to design programs which manipulate words strictly according to the grammatical rules, it has been much less easy to give them any idea of what the words mean. In part, this difficulty is a consequence of the other two factors discussed above, for meaning is obviously something which is very hard to grasp except through experience, and the knowledge of meaning that is gained through experience must be generalized if it is to be employed in language. Again, we shall come back to this subject in the next chapter.

There are, however, circumstances in which we ourselves do make use of the sort of logical model of deductive thinking with which AI finds easiest to cope. Disciplines like medicine and the natural sciences have advanced as a result of our growing ability to reduce the apparently arbitrary and random behaviour of nature to a systematic and orderly series of rules and laws. This process could be viewed as one of making a model of nature which, by extracting underlying generalities, avoids the combinatorial nightmares which would arise if, for example, a doctor had to treat every case of 'flu as a unique phenomenon or a chemist had to explain every reaction anew in terms of first principles.

Scientific data which has been codified and organized in this way fits very easily into artificial models of intelligence, for the process of establishing and proving a scientific hypothesis is, in effect, that of transforming heuristics into firm principles, or of pruning back the colossal decision tree which might link possible causes with discernible effects to the hardwood which can survive the test of repeatable experiments. To draw an analogy, one might describe science as an attempt to discover the rules which nature follows in moving its chess-men around the board. Once these rules have been established they can be fitted fairly easily into a game playing model of intelligence.

If knowledge of this kind is threatened by a combinatorial explosion, it is one which is a consequence of the very success of science; for as specializations multiply and diverge the number of boards in use, the variety of pieces, or facts, in play and the systems of rules, or natural laws, that apply are in danger of outstripping the human capacity to absorb and retain information.

Here artificial intelligence does have something to offer in the shape of the 'expert system'. The computer's ability to assimilate, store and accurately retrieve information surpassed our own at a very early stage, as soon as reliable back up memories had been invented. The expert system is a method of taking advantage of this colossal storage capacity in order to collect and collate the knowledge of a number of human experts in a particular field and thus create an electronic expertise which can provide, on demand, a range of knowledge and experience greater than any individual could possibly accumulate in a lifetime's experience or retain in his memory.

One of the fields in which expert systems have proved particularly useful is that of medicine; for it is clearly of great value to a doctor to be able instantly to obtain a second, specialist opinion based on the most authoritative and up-to-date information. But they are also used in subjects as diverse as geology and law and versions are under development which can be sold, in software form, for use on mini and even microcomputers in the home.

The kind of intelligence that expert systems employ is very closely related to traditional, textbook logic of the kind which deals with propositions and their implications. For example, the logic needed to derive diagnoses from symptoms is exactly the same as that employed in familiar examples such as: All dogs have four legs. All dogs bark. Fido has four legs and barks. Is Fido a dog? Recently, the invention of a computer language, PROLOG, which allows all information supplied to an expert system to be formulated as implicative statements which the program can manipulate according to the rules of predicate calculus, has produced, enthusiasts claim, a framework which will allow any codified body of knowledge to be assimilated into an expert system and retrieved quickly and flexibly.

A good medical system will respond, for example, when supplied with a list of the symptoms which a patient exhibits, with not only a list of those diseases or conditions which might have given rise to the symptoms, but also with suggestions about further symptoms which might be sought in order to confirm a particular diagnosis. A further refinement, of great importance when expert systems are applied in a field like medicine, where an error could have disastrous results, is that the program can be instructed to explain its own reasoning. The user can thus check that the logic the program has employed is watertight and that it has not made a mistake due to some oversight in its own data base or as a result of being supplied with incomplete information by the user. It may be obvious to a human doctor that the patient who complains of nausea and a craving for unlikely foods is female and of childbearing age, but an expert system has no way of knowing this and unless it has been told might well conclude that the symptoms indicated acute appendicitis.

Whether expert systems can be said to be truly 'intelligent' is very much an open question. Proponents argue that it is possible to devise programs which, given input from human experts that is formulated with sufficient subtlety, can distinguish and clarify ideas which the individuals concerned might think of as merely suppositions or 'hunches' – though it must be debatable whether it is really desirable to undertake major surgery or drill an oil well on the basis of a computerized hunch. It would seem clear, on the other hand, that the real intelligence which is involved is that of the researchers who provide the raw information the system uses, and that only if an expert system went out into the world and collected its own information would it be in a position to lay claim to a comparable standard of intelligence. But any attempt to produce an expert system which did its own research would, of course, run head on into the problems of the combinatorial explosion.

Our own conclusion is that the expert system, however useful it may prove as a means of indexing and cross-referencing the ever-expanding corpus of scientific information, is basically no more than another manifestation of the computer's well-established talent for storing and manipulating symbols according to logical rules: as such, it does not provide us with a model of intelligence that is particularly illuminating. After all, even for our own purposes, we do not accept information as genuine expertise, and therefore proper fodder for an expert system, unless it is formally set out according to the same logical systems as those which expert systems employ and unless it has been firmly fitted into place on a clearly defined and well-established decision tree. An expert system 'thinks' only in the formal, articulate, rational mode we ourselves use when writing a book or preparing a scientific paper. If it is a model of intelligence, then it is a model of a kind of intelligence which we already understand rather well.

It can be argued that expert systems, and the game playing and problem solving models, are in reality only models and that if we grant them any validity we are misunderstanding the nature of our own intelligence. What is at stake is essentially a philosophical issue which centres on what we mean when we say we understand something. Understanding is, after all, something that can take place at different levels: an infant understands, simply as a matter of common experience, that if objects are dropped they fall to earth; the same child may subsequently learn that this phenomenon is but one manifestation of the force of gravitational attraction which can be expressed in a series of mathematical equations; later still, if the adolescent pursues a scientific bent, he will discover that, in terms of fundamental physics, no one fully understands gravity at all! But the most important distinction is between those things we understand as a result of our own experience, which we have learned at first hand, so to speak, and those things which we understand because they have been taught or explained to us in the abstract. The growth of human understanding can be seen as the process by which we extrapolate the first kind of understanding into the second. Thus from the observations that objects fall and that the planets stay in orbit, Newton was able to formulate the concept that objects attract each other.

The difficulty with artificial intelligence, critics suggest, is that it relies entirely upon understanding of the second kind. Thus it is possible to devise a chess playing program which has considerable understanding of the relationship between chess-men as they are moved about the board according to formal, abstract rules which can be expressed in symbols, but, so far, impossible to produce one which can understand the same assortment of chess-men when they are lying higgledy-piggledy in a box, where any attempt to express their relative positions in symbolic, algorithmic terms collapses in the face of the combinatorial explosion.

Human beings are of course accustomed to thinking of the situation on the chess-board as requiring infinitely more intelligence than that in the box, but this assumption is only made because we take almost entirely for granted the intelligence which we have acquired through experience, even though, in information processing terms, it may be far more demanding than that needed to play a decent game of chess. A child quickly learns that a pyramid will not remain upright for long when balanced on its apex, and extrapolates the experience thus gained into a piece of commonsense knowledge: that heavy objects quickly topple one way or the other if one tries to balance them in unstable positions. A program like SHRDLU, which manipulates shapes, will also 'know' that a pyramid cannot be 'placed' on its point, but only because it has been given a set of rules, including one which states that a pyramid must be put down on its base. The program contains no 'intelligence' that would see anything intrinsically absurd in building the Great Pyramid upside down.

The issue of the level of understanding which is achieved by AI programs arises even more directly in the case of programs dealing with vision and language which are the subject of the next chapter; but it is clear, even in the game playing or problem solving field, that the difficulties are at base due to the kinds of models that are being employed. All artificial intelligence work starts from what has become known as the 'top-down' approach; it tries to provide machines with programs that encompass relatively sophisticated kinds of intelligence. This effort to provide pre-packaged programs, which contain all the 'intelligence' a machine will need to play chess or diagnose an illness, has begun to come under fire from those who argue that real intelligence will only be achieved by quite different means, working from the 'bottom-up': the failures of AI, it is argued, are essentially a consequence of trying to make machines run before they can walk.

The debate is not simply concerned with what kind of tasks should be attempted, but with the fundamental question of which level of information processing should be taken as a basis for intelligence. As we have seen, the classic, top-down approach of artificial intelligence has taken the 'higher faculties' of reasoning, deduction and abstract thought and attempted to extract programmable models from them. The failure of this approach to produce a model of intelligence which can be applied to the real world has led to the suggestion that the only possible alternative is one which starts with what we actually know about the basic information processing components of the brain, the networks of neurons, and build intelligence from the bottom up. Only then, it is argued, will we be able to act in the confidence that the artificial intelligence we are examining is really the same thing as natural intelligence and not just a superficial model which looks deceptively similar under some

Like many discussions involving two sides which start from irreconcilable points of view, the controversy between 'top-downers' and 'bottom-uppers' can quickly degenerate into fruitlessness. Against the success of the top-down approach in fields like game playing and expert systems, the bottom-down enthusiasts can balance their achievement in pinpointing the areas of the human brain involved in other kinds of intelligent behaviour. For example, the British physiologist David Marr was able to show that it is a particular section of the brain, the cerebellum, which translates an instantaneous decision, such as determining that the response to a careless return in tennis should be a rush to the net with a forehand volley, into muscular action. In circumstances of this kind, the brain does not behave like an AI program, planning a sequence of limb movements from first principles, but the decision to run forward and play forehand is implemented by drawing on one out of a whole set of responses which are stored ready for immediate use. From Marr's work one might therefore draw the conclusion that, if it is wished to discover more about the way in which the brain organizes physical movement of the limbs, the cerebellum is the place to start one's search.

But those who support the top-down approach can counterattack, when its shortcomings are criticized, by ridiculing the idea that it would be possible to explain the difference between a good tennis (or chess) player and a bad one by painstakingly analysing the characteristics of hundreds of thousands of neurons.

The controversy is, of course, of enormous importance to robotics. Engineers must choose, in designing the mechanical intelligences of future robots, between the bottom-up school, urging that a sound foundation can only be established by looking to the anatomist and the neurophysiologist for such clues as they can offer about the natural brain, and the top-down school which takes the digital computer as given and argues that all difficulties will, in the end, succumb to the skill of programmers.

Our own conviction is that success will be achieved by combining insights from both schools of thought. As we will show in Chapters Ten and Eleven, a good deal of intelligence can be imparted to machines built on the principle of a neural net, with inherent, unprogrammed learning ability. But that intelligence can be refined and extended by the use of ideas borrowed from the top-down approach. There does seem little doubt, however, that in setting its sights so resolutely on the high peaks of human intelligence, AI has overlooked whole ranges of foothills which must first be scaled. In particular, game playing and problem solving models will never achieve value in the real world unless the basic problems of perception and the interpretation of sense data and language are first solved. But, at the same time, the solution of these and other difficulties will certainly be aided by ideas and techniques drawn from the top-down approach.

9. Eyes to See and Ears to Hear

The ring where name and image meet.

W H Auden, 'Prologue'

The mastery of vision and language is very high on the list of challenges which robotics must overcome if robots are to become truly adaptive and versatile. Industrial robots will have to be able to see what they are doing if they are ever to break out of the sightless world of preprogrammed repetition in which they currently exist; and if they are to work alongside human beings, the potential for co-operation will remain strictly limited until and unless communication can be achieved in everyday language and without the use of keyboards and specialized vocabularies. Slightly less obvious, perhaps, is the importance of sight and speech to the more general problem of creating a significant level of mechanical intelligence.

It is not only the combinatorial explosion which has restricted AI to the world of the game board and the playpen; it is also the fact that, lacking the ability to see the world around them or to understand natural language, machines have a limited range of topics to which their intelligence can be applied. In practice, sight and language go hand in hand, for understanding what is seen inevitably involves attaching labels or descriptions to it, and unless a machine can see for itself, it will have little use for language – what could it talk about? The interdependence of the two faculties is in fact very evident in the case of conversation between two human beings. If people talk to one another, then they must either talk about something that both have seen or discuss a subject about which both have acquired information by reading or listening. The only exception to this rule is the teaching process, where one person conveys information that is novel to the other; but in practice even this is reliant on some shared basis of experience – how can one describe a sunset or a landscape to someone who has been blind from birth?

The first person to tackle the problems of relating vision and language seriously was Terry Winograd, whose SHRDLU program, developed in the early 1970s, was intended to enable a computer to see and understand a tiny world of cubes and pyramids and also to describe to a human interrogator what it saw and how it would go about manipulating the objects in order to achieve prescribed objectives. As we saw in the previous chapter, SHRDLU relies upon a classic 'top-down' artificial intelligence approach to manipulating its world: the program does not 'understand' that cubes are three-dimensional or that pyramids can rest stably only on their flat bases, but operates within the rules it has been given exactly as a chess player operates within the rules of the game. The techniques that Winograd and others have used to tackle the problems of vision also start from the traditional 'top-down' paradigm. The task they set themselves is generally described as 'scene analysis' and their methods are based on the assumption that it should be possible to devise sets of rules or algorithms which will enable a computer to interpret a visual image by breaking it down into a geometrical pattern which can then be identified by reference to a store of sample patterns in its memory.

It must be emphasized at this point that scene analysis is not the only possible way to approach the problem. We shall be looking in subsequent chapters at the rather different methods which are generally categorized as 'pattern recognition' and which may not involve preprogramming or the formulation of algorithms, but rely instead upon the ability of learning automata to construct their own state structures in response to inputs. In this chapter, however, we will be concerned with the progress that the programming approach has made, initially with vision systems and then with natural language.

The possibility of artificial or robot vision was first mooted towards the end of the 1950s when the American space agency, NASA, was contemplating the chances of landing robot vehicles on the moon and the inner planets in order to explore their surfaces. Such robot explorers would obviously be far more useful if they could see where they were going and interpret what they saw in order to avoid obstacles, collect samples and, not least, find their way back to base. The fact that when, in the 1970s, unmanned vehicles did land on planetary surfaces they had to relay pictures back to Earth in order to obtain instructions, or simply operated blindly, reflects the enormous difficulties that vision presents to artificial intelligence.

Indeed, when NASA first expressed interest in the subject scientists at the main centres of AI research, MIT, Stanford and Edinburgh universities, scarcely knew where to start. The scale of the problem can be appreciated if we look at the sheer amount of information that has to be processed before any intelligence, as we would define it, can be brought to bear. The actual 'capture' of an image presents little difficulty; a television camera not only collects all the necessary information about a scene which is presented before its lens, it can also communicate it in convenient, digital form. As is reasonably well known, the picture on a television screen is made up of rows of dots, or pixels, and the electron gun in the cathode ray tube scans the screen, moving horizontally back and forth along the rows of pixels, 25 times a second, refreshing or changing the intensity of each pixel in turn. A picture with the quality of the ordinary domestic television set will contain about 250,000 pixels; and the intensity of light in each pixel can, typically, be varied on a scale from 0 to 255 in order to convey tones between pure black and pure white. Each pixel can, therefore, be represented by an eight digit number in the binary code (between 00000000 and 11111111) and the processing of the complete image involves a quarter of a million such numbers, or two million bits of information in total. It is only comparatively recently that it has become possible, as a result of the invention of a device called the intelligent framestore, to perform this feat in 'real time' – to process 25 separate images each second as fast as the camera generates them. The actual operation of the intelligent framestore need not concern us greatly; its function is to store, momentarily, an entire image and to operate on it intelligently in order to identify and extract its significant features.

The important question is, of course, what features of an image are significant? In order to tackle this question the pioneers in the field had first to simplify the scenes which were to be analysed. The surface of the moon, Mars or Venus might well present images which would be puzzling to a human eye, let alone that of a rather primitive machine, and it is, of course, a prerequisite of the AI approach to vision that the programmer anticipates every possible object that will be seen since he must devise an algorithm for recognizing it. Researchers in the field of vision systems, like those concerned with problem solving, were thus forced to reduce the content of the worlds in which their machines operated to simple objects like children's building bricks. At this level a possible line of attack became clear.

If lit appropriately, so that confusing shadows were eliminated and the corners between surfaces on different planes emphasized, it was feasible to extract from a scene containing a number of simple, angular shapes a 'line drawing', the sort of thing that a human might produce by laying a sheet of transparent paper over a photograph of the New York skyline and tracing the shapes of the prominent buildings. In such a picture the significant features are, obviously, the lines which mark the edges of objects and the corners where two or more edges meet. Simply extracting simplified line drawings of this kind is a formidable task and one which, until the arrival of the intelligent framestore, involved a great deal of sophisticated programming.

This feat of extracting significant features from an image is, however, something that we ourselves must have mastered at some stage. For there is a good deal of evidence to show that the role of the visual cortex, the section of the brain which initially receives the information transmitted from the rods and cones within the retina of the eye, is to concentrate on areas of an image where sudden changes of intensity or tone mark angles and edges. In other words, it picks out and stores the essential information, consisting of a few lines which compose a sort of lightning sketch of what the eye is seeing, and the rest of the brain can use this sketch to 'get its bearings' on a particular image. It seems probable that it is because we habitually 'preprocess' the data from the eye in this way and quickly identify the main features of it, that we so easily recognize a picture in which a complex image has been reduced to a few significant lines. It is difficult, otherwise, to account for the ease with which we instantly recognize a cartoonist's image of a politician or reconstruct a familiar face from the few, quick strokes of the pen which a portrait artist may use to indicate the hairline, the shape of the nose and the line of the jaw.

With artificial systems it is necessary to allow some flexibility, so that missing bits of a line, perhaps obscured by reflected light or camera 'noise', can be filled in to yield a 'sensible' image. But the real difficulties start once the line drawing has been obtained – how is a computer to go about the work of translating the collection of digits to which the picture has now been converted into a realization that what is being seen is, in fact, two cubes and a pyramid, or whatever? As always, with a preprogrammed system, the need is for an algorithm.

The first real progress towards devising an appropriate way of making the problem 'computable' was made by R B Roberts at MIT, and relied heavily on the rules of projective geometry, the sort of laws that we all master when we first appreciate that a cube can be represented in three dimensions rather than only two. The elements of Roberts' methods can be illustrated if we take a simple example of the kind of scene which AI has successfully analysed. There are, let us say, two objects in view, a cube and a wedge shape. Once the lines marking the edges of the various planes have been found, the corners where they meet can be identified and numbered, thus:

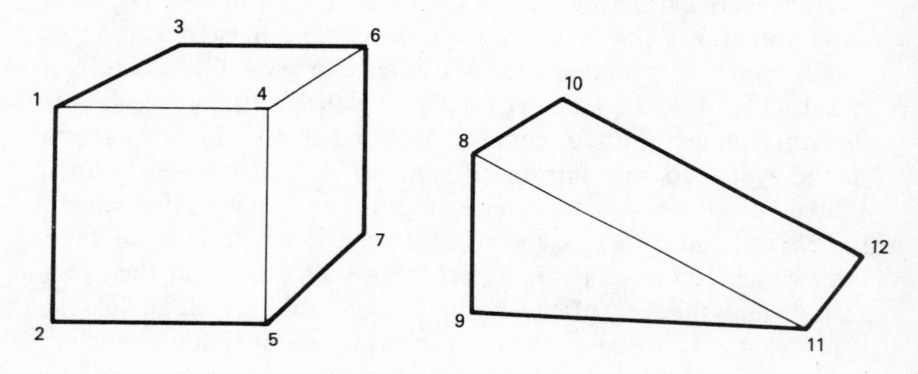

The scene with cube and wedge

The relationship between the corners, that is their position within the frame of the image and the way in which they are joined by the edges of the objects, can now be incorporated in a matrix. For the sake of clarity we will show this diagrammatically, though within the computer's electronic circuitry it will obviously take the form of a sequence of digits. If we list the corners along both axes and place a 1 on the grid wherever two corners are joined by an edge, the result will be a matrix like that shown opposite.

It is immediately clear that, since they have no edges in common, the two groups of corners, 1-7 and 8-12, represent two separate objects. It should also be evident that, leaving aside the improbable circumstance where an object is seen head on, any six-sided shape like the cube will have seven corners and form a 7 × 7 matrix and that a wedge shape will have four corners except when three of its surfaces are visible, when the number of corners and the matrix will both be larger.

	1	2	3	4	5	6	7	8	9	10	11	12
1	0	1	1	1	0	0	0	0	0	0	0	0
2	1	0	0	0	1	0	0	0	0	0	0	0
3	1	0	0	0	0	1	0	0	0	0	0	0
4	1	0	0	0	1	1	0	0	0	0	0	0
5	0	1	0	1	0	0	1	0	0	0	0	0
6	0	0	1	1	0	0	1	0	0	0	0	0
7	0	0	0	0	1	1	0	0	0	0	0	0
8	0	0	0	0	0	0	0	0	1	1	1	0
9	0	0	0	0	0	0	0	1	0	0	1	0
10	0	0	0	0	0	0	0	1	0	0	0	1
11	0	0	0	0	0	0	0	1	1	0	0	1
12	0	0	0	0	0	0	0	0	0	1	1	0

Matrix representing the cube and wedge

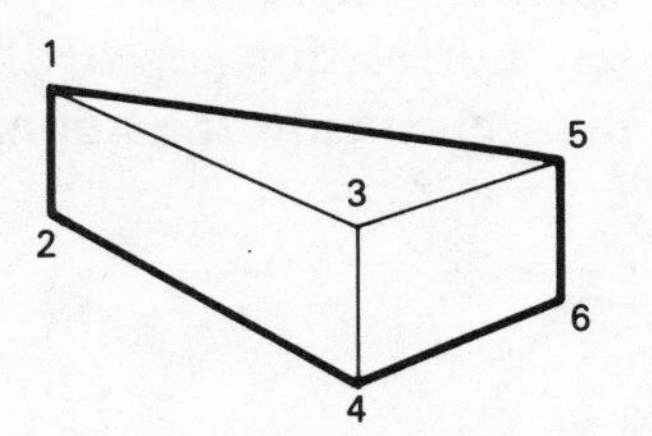

	1	2	3	4	5	6
1	0	1	1	0	1	0
2	1	0	0	1	0	0
3	1	0	0	1	1	0
4	0	1	1	0	0	1
5	1	0	1	0	0	1
6	0	0	0	1	1	0

A wedge with three sides visible and the corresponding matrix

It is relatively easy to see that a program could be constructed which stored a variety of matrices, each corresponding to an object of a particular shape, and which could also relate the orientation of an object to a matrix of a particular form. If a few further rules are added the program can also cope with more elaborate scenes. Suppose, for example, that the wedge is moved so that, from the camera angle adopted, it overlaps the cube. The corners in the scene would now be numbered as follows.

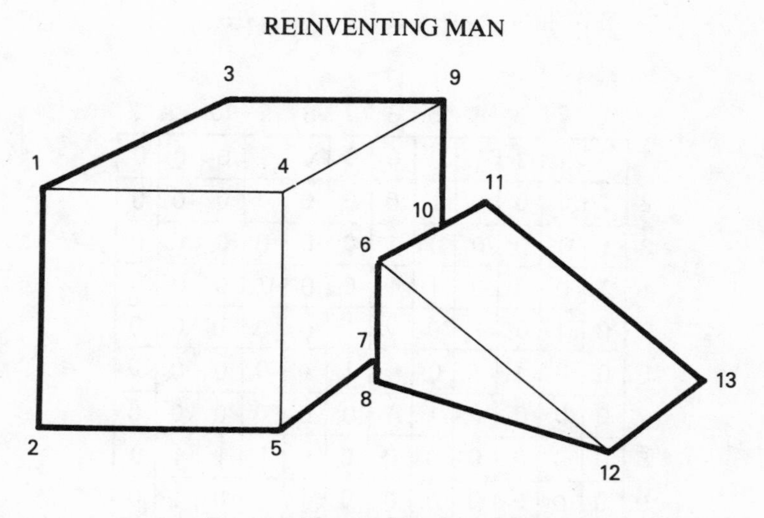

Scene with overlapping objects

Providing that the program included a rule that 'T-junctions' such as those at corners 10 and 7, which only occur when two objects overlap, should be ignored, it would still come up with a matrix that showed two separate objects and although the bottom right hand corner of the cube is now obscured, the matrix for the cube would still be recognizable as such if the program was allowed a little leeway to cope with this sort of contingency. The addition of further rules, perhaps those governing the interpretation of parallel lines, would allow the program to distinguish between objects of a different shape, even though the number and relationship of the corners were the same. For example, in the scene below the sloping

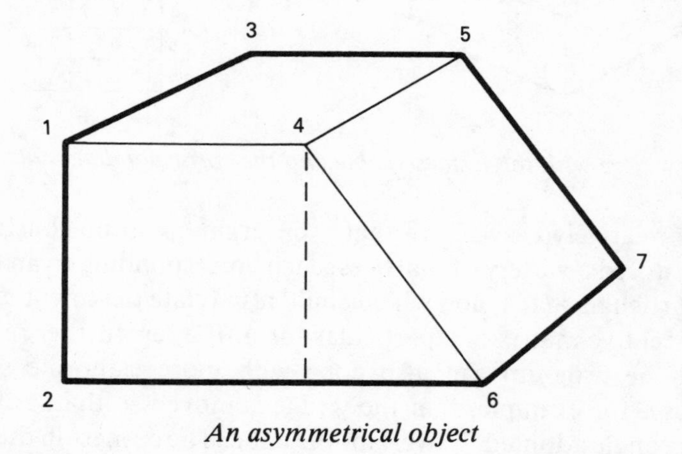

An asymmetrical object

lines joining corners 4 and 6 and 5 and 7 would betray the fact that the object was not a cube because they are not parallel with those joining corners 1 and 2.

It is easy to see, however, that even if restricted to a world containing only the regular polyhedra (that is, the shapes of classical geometry which are, perhaps, rather commoner in the classroom than in the factory, let alone on the surface of the moon) Roberts' methods quickly come to involve computational processes of nightmarish size. Several researchers therefore began to look around for algorithms which might be simpler to apply. The approach which has proved most significant and valuable was that pioneered by Adolfo Guzman at MIT and later elaborated by Max Clowes at the University of Sussex. Again, the starting point is the simplified line drawing of the scene, but having found the edges and corners, the programs devised by Guzman and Clowes use them as clues to the relationships between the surfaces which they enclose. Thus, going back to the first scene we examined, and numbering the corners in the same fashion, we can go on to identify, by letters, five surfaces.

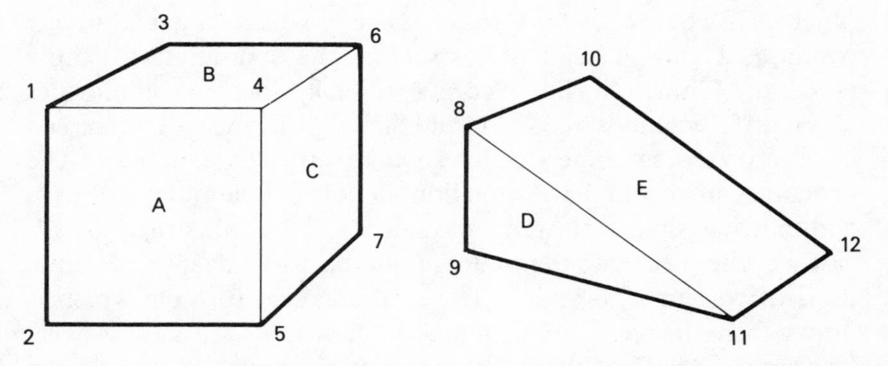

The scene with cube and wedge showing the five surfaces labelled

Ignoring for the moment the geometrical implications of the corners, it is possible to formulate a mathematical description of the relationship of surfaces and corners which can be shown diagrammatically as overleaf.

Here the circles, or nodes, represent the surfaces and the connecting lines are numbered to indicate which corners they have in common. Again, it is possible to deduce that two separate objects

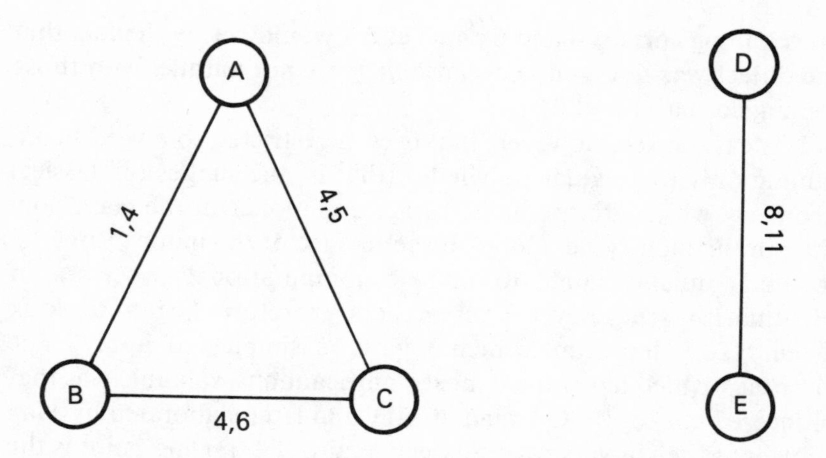

Graph representing the relationship between surfaces and corners

are involved and it is also clear that objects with different shapes will generate diagrams or, within the computer, formulas, of a different kind.

The Guzman-Clowes technique can also accommodate rather more sophisticated ideas about the relationships between different objects. If we now go back to the scene in which cube and wedge overlapped and, instead of instructing the computer to simply ignore the T-junction kind of corner, equip it with some simple rules about different kinds of corners and their significance, the scene can be interpreted in some detail. To take just two examples: The program can be told that T-junctions denote overlapping of objects and that star-shaped conjunctions, such as occur at corner 4, indicate a corner nearer to the camera than the arrow-shaped conjunctions at corners 1, 5 and 6. The mathematical formula which is drawn from the scene will then make it clear that the wedge is overlapping surface C of the cube and that the camera is viewing the cube obliquely from the top right. It has been possible to extend algorithms of this kind to deal with curved and shadowed surfaces and more complex shapes. But it would be simply impractical, even if it was in principle possible, to provide a computer with a memory of the formula which results from every possible object that might come into view in even a relatively restricted world. For most real-life objects have composite shapes: if asked to describe something as simple as a bolt and nut in terms of their geometrical forms, we would have to say that the first was a rod with a disc on one end and the latter a thick disc with a hole through its centre. There is in fact a good deal of evidence to support the belief that we ourselves think of complex objects in this way, as combinations of simpler forms, at least when, as children, we are first learning about them. Small children, for example, usually draw trees as upright pillars with fairly standardized fluffy leaves on top, or people as globular heads and torsos with simple, matchstick limbs attached.

The key step that advanced AI from simple to composite shapes resulted from the work of P H Winston of MIT in the mid-1970s. Given that the methods invented by Guzman and Clowes had by this point achieved a degree of reliability in identifying a range of simple objects, Winston showed how, once it had identified them, a program could arrange these elements in what he called 'semantic nets' in order to represent more complicated objects. Taking that favourite philosophical example, the table, we can see how its constituent parts combine into one of Winston's nets. A proto-typical table could be described as four pillars supporting a slab, in diagrammatic form, something like this.

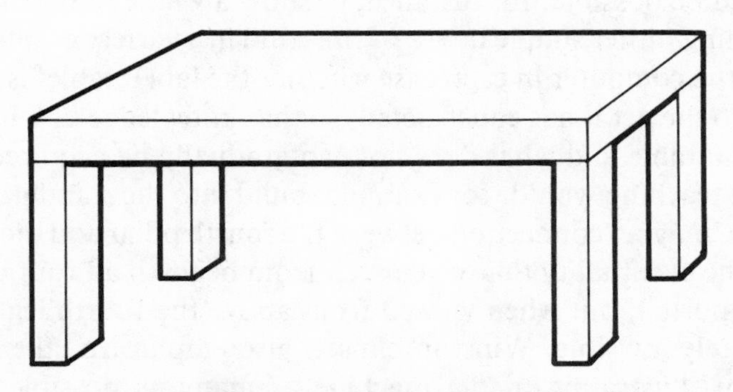

A table consisting of four pillars and a slab

It is fairly simple to see how the Guzman-Clowes technique would allow a machine to identify four (or, if the perspective was slightly different, three) pillars and a slab. The rules about the interpretation of T-junctions would also yield the information that, in the case of the three pillars which are visible in their entirety, the corners of the slab overlap the top of the pillars, which could convey the message of 'corner support' (CS) for the slab. A program based on Winston's methods would then be able to construct a semantic net for the image which could be represented, in diagrammatic form, like this:

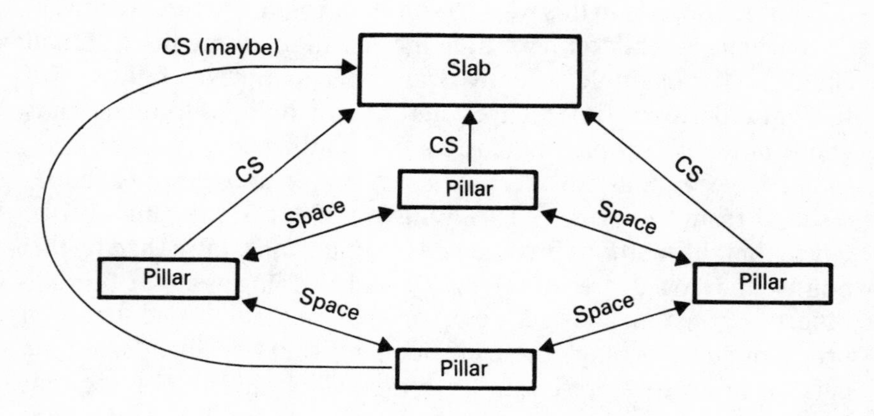

A semantic net for a table

A program which is able to construct semantic nets of this kind can learn what they mean, or represent, through a teaching process. It would be possible, for instance, to show a whole range of scenes containing rather simple tables of this kind in a variety of positions, telling the computer in each case whether the label 'table' is appropriate to the net it has constructed, so that knowledge of what constitutes a table and what does not can gradually be acquired. This kind of teaching would, for example, build into the standard 'table net' the 'maybe' connection between the fourth pillar and the slab to cover the eventuality that when seen from beneath all four corners are supported, but when viewed from above the fourth leg will be completely invisible. Winston himself also emphasized the importance of illustrating to the machine as many as possible of the objects which were almost but not quite 'tables' – see the 'non-table' shown opposite, for instance.

A computer using programs of the Guzman-Clowes and Winston type is thus capable of learning by experience to some extent, and is capable of some degree of generalization. It might well, for example, correctly identify a table of a kind it had never seen before. Unfortunately, its usefulness is limited by the fact that most problems connected with vision do not take the form of distinguishing a table from, say, a chair, but demand that one table be distinguished from another. Though semantic nets learn inductively, they cannot

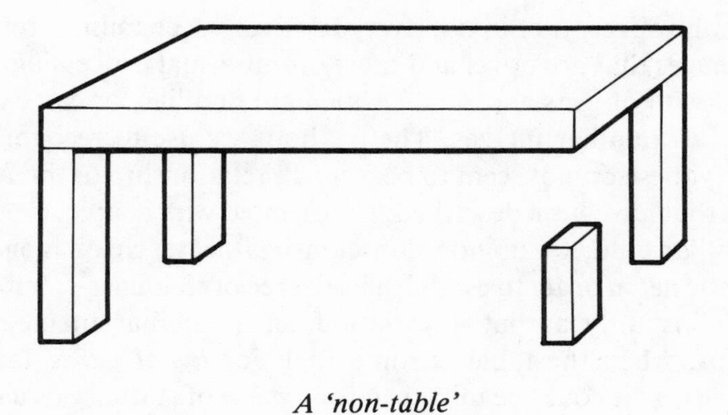

A 'non-table'

easily master the equally important task of recognizing particular instances of an image rather than its generalized conception. This difficulty, as we shall see in the next chapter, can be overcome by unprogrammed pattern recognition systems which are also capable of achieving generalization without being given, or constructing, algorithms. This enables a pattern recognizer to store and recognize images which have not been foreseen by its designer. A preprogrammed machine using semantic nets, on the other hand, can only cope with the real world, or even small, carefully planned bits of it like a workbench or an assembly line, if the programmer anticipates the constituent parts of every kind of image that may be encountered and provides algorithms that enable the machine to interpret it correctly. For although the program may be able to construct a semantic net for a novel composite object, it can only do so if it has been programmed to identify all the bits that go to make up the object. Rather clearly, such systems very quickly run up against the problems of the combinatorial explosion and, again, their inability to provide the adaptability which seems to be built into the human system leads one to question whether they are based upon an accurate model of its workings.

The idea of semantic nets certainly strikes some chords when we consider our own perceptive processes. When we encounter an unfamiliar object, we do tend to analyse it and try to describe it in terms of its component parts, though our references are drawn from a much wider selection than any computer program employs. Faced with the task of describing a novel piece of architecture, for example, we might do so by reference to familiar shapes – talking of an 'umbrella-like' canopy, say, or a 'mushroom-shaped' dome. The fact is, however, that in our everyday lives we encounter relatively few images that are novel and most of our visual processing is concerned with objects and scenes which are familiar, or can easily be related to familiar images. The methods we use to recognize and identify these do not seem to bear much relationship to the AI programs that have been described. Confronted with a table, even of an unfamiliar kind, we do not (consciously at least) check it against a semantic net in order to establish its degree of 'tableness', but recognize it instantly as but a variation on a familiar theme. Only, perhaps, when the table is some flight of *avant garde* fancy so extreme that it could be mistaken for a piece of statuary or a house-plant holder do we actually refer to an algorithmic description of what constitutes a table in order to decide whether we can legitimately use it as a resting place for a cup of coffee.

The likelihood that the preprogrammed algorithmic approach is inadequate must be reinforced by the fact that, in practice, it has been quite unable to break out of the simplified world of stylized objects into the real world of visual complexity. Where preprogrammed visual systems have found a use in automation or robotics they have been limited to applications in which the world with which they are asked to cope is of a guaranteed, elementary kind – the kind, in fact, where in many cases automation has already devised cheaper and less demanding techniques.

The alternative approach, that of adaptive pattern recognition, may offer a better prospect for the future, if only because the ability to recognize patterns, even very complex and amorphous ones, seems to be the fundamental visual skill in both human beings and other animals. But pattern recognition is even less susceptible to top-down, preprogrammed techniques than scene analysis. It does, however, yield to mechanical methods surprisingly easily once we replace the notion of a man-made intelligent program with that of a machine which has the capacity to *become* intelligent as a result of a learning process.

We shall try to show, in the next two chapters, that although the programming techniques developed by AI certainly have an important role to play in visual systems, it seems likely that they will only realize their full potential when they work in concert with the very different methods of adaptive pattern recognition systems.

If we now turn from vision to language, we find that virtually all the work that has been done to date again relies upon preprogramming. On the face of it, it might seem that language is a field in which computers had a flying start. For it was evident from the very first that, if the digital computer was to be a practical tool, it would have to use a language of some kind; the prospect of reducing all programs and data to the stream of binary digits which is a computer's 'native' language was both tedious and impractical. Moreover, the translation of simple language into the binary code or vice versa is a mechanical task well within the machine's own powers.

It is not, however, the case that any existing computer language provides a basis on which machines can build an understanding of natural language as we use it ourselves. A computer can be said to understand the language it uses in only a very limited sense. For example, a machine which uses BASIC can be told to 'PRINT' or 'LIST' and will, in response, print out a result or list a program; but it is reacting to the instruction as a dog will obey an order to 'sit' or 'stay'. Machine and animal have both been programmed to react in a certain way on hearing a given word, but the computer no more understands the meaning of the word 'print' than the dog understands the word 'sit'. One could program the computer to print when it was told to 'sit' as easily as one could teach a dog to sit when ordered to 'print'.

The difficulty with natural language is that, if it is to have any meaning, it must be *about* something, and given that a computer has no knowledge beyond the information that is contained in its programs, there is no starting point for teaching it what words mean. This problem can perhaps be made more explicit if we imagine the difficulties of trying to teach a human student a language in circumstances where the teacher is forbidden to translate any of the vocabulary or to indicate meaning by gesture or example. In such a case, an assiduous student with an impeccable memory might well accumulate a large vocabulary, and might also master the grammatical rules, the conjugation of verbs, the conventions governing the use of capital letters and so forth, but though he might subsequently utter sentences of impeccable style and syntax, the fact would remain that there is no way in which he could attach any meaning to what he spoke or heard.

The pointlessness of learning a language in this way becomes very plain in the case of computer translation from one language to another. Translation would seem, on the face of it, a chore which the computer should master without difficulty; it is, after all, easy enough to store the contents of a bilingual dictionary in a computer's memory and to program the business of looking up translations and organizing word order, etc. But a computer with a program of this kind is in the position of a pupil who has learned *two* previously unknown languages, and the rules for translating one into the other, without learning how to translate either into his native language. If it is asked to translate, say, three French phrases such as 'le film bleu', 'le biftek bleu' and 'le ciel bleu', it can only be a matter of chance as to whether the film, the steak and the sky emerged in translation, respectively, as 'pornographic', 'undercooked', and 'azure' – they might equally well all turn out as 'melancholy'. It is because meaning is so elusive when unanchored by such experience that attempts to get computers to create sentences in natural languages lead very easily to faultlessly grammatical but hilariously nonsensical 'computerspeak' such as: 'High pink theories walk gratefully in the kitchen.'

A limit can be put to this sort of fatuity, of course, by supplying further rules; to the effect, for example, that physical qualities cannot be attributed to abstract nouns. But though this may ensure that computers appear more cogent, it does not solve the basic difficulty of enabling them to know, or learn, what they are talking about.

The only possible solution would seem to be to couple the acquisition of language to a program which also perceives or acts upon the world which is to be talked about. This was, of course, one purpose of Winograd's SHRDLU program, and his pioneering work has subsequently been followed up by a number of others. In order to see how AI set about the task of equipping computers with even very elementary forms of natural language, we can take a small, strictly circumscribed world of the kind that is used by the robot task programs discussed in the previous chapter. Assume, for example, that, equipped with a progam for scene analysis, a robot has learned to interpret the various scenes that can be made up from a selection of nuts and bolts. There are six such objects in view at the time in question, and each kind of object can appear in two varieties, large and small. The nuts and bolts each occupy one of the eight fixed positions within the image, numbered 1-8, and each object is labelled with a letter from A-F.

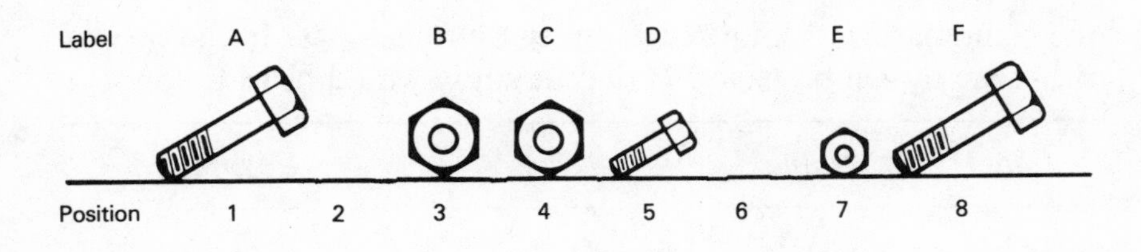

A robot task scene

The scene consists of two large bolts, A and F, two large nuts, B and C, one small bolt, D, and one small nut, E. The vocabulary will consist of the definite and indefinite *articles*, two *nouns*, 'nut' and 'bolt', two *adjectives*, 'large' and 'small', three *verbs*, 'is', 'pick' and 'place' and two *relation* phrases, 'to the left of' and 'to the right of'.

The grammar, which is the heart of the program, dictates that a noun phrase is made up of an article and adjective and a noun and that three kinds of sentences will be used by the person operating the computer: questions, assertions and commands. The form each of these takes is also specified, thus:

Question = Is (noun phrase) (relation) (noun phrase)?

Assertion = (noun phrase) is (relation) (noun phrase).

Command = Pick (noun phrase) *or* Pick (noun phrase) (relation) (noun phrase) *or* Place (noun phrase) (relation) (noun phrase).

Given this grammar, the program can parse any sentence which is input (that is, break it down into its constituent grammatical parts) in order to ascertain which category it is in and whether it is within the rules and, therefore, comprehensible to it. For instance, it will parse the question 'Is a small bolt to the left of the large nut?' and establish that it is made up of 'is' followed by a noun phrase, a relation phrase and another noun phrase and that it therefore must be a question, rather than a command or an assertion, and that it is in the proper form. The parsing process relies upon the program constructing a data base, that is a catalogue of all the information it has about the image presented to it, and then manipulating that information according to logical rules.

Suppose that the scene analysis program has identified the objects and their positions in a particular image; the language program can then summarize its knowledge under four headings. In the case of the scene shown on page 195 the summary would be as follows:

Label	*Name*	*Size*	*Position*
A	Bolt	Large	1
B	Nut	Large	3
C	Nut	Large	4
D	Bolt	Small	5
E	Nut	Small	7
F	Bolt	Large	8

The program can then classify every one of the labels in terms of its vocabulary, thus:

The nouns are (ABCDEF).

The labels which denote nuts are: (BCE).

The labels which denote bolts are: (ADF).

The labels which denote small objects are: (DE).

The labels which denote large objects are: (ABCF).

Each of these groups is a mathematical set, the set of all nouns, the set of all bolts, etc and the relationships between their positions can also be expressed as sets: for example, the set of all objects to the right of (C) is (DEF). With all the bits of the image categorized in this way, the use of elementary logic allows the program to find the label for any object which is described by the vocabulary. For example, 'small bolt' is D, the label that is common to the set of bolts (ADF) and the set of small objects (DE). The large bolt to the right of the small nut is F, the label that is common to the set of bolts (ADF), the set of large objects (ABCF) and the set of objects to the right of the small bolt (EF).

The vocabulary that the computer employs to communicate with the operator is very straightforward. Assuming that the grammar of the input sentences is correct, it will answer questions with a 'no' or a 'yes' and a label (A, B, C etc), statements with 'true' or 'false', and will obey a command if it is possible. Suppose that a question is put: 'Is a small nut to the right of the small bolt?'. The program would translate this into labels as follows:

Is a (DE) (the set of small objects) (BCE) (the set of all nuts) to the right of the (DE) (the set of all small objects again) (ADF) (the set of all bolts)?

Eliminating the labels which are not shared by the nouns and adjectives in each noun phrase simplifies this to:

Is (E) to the right of (D)?

It is now possible for the program to identify (by referring to the list of numbered positions) the set of objects to the right of D as (EF) and so the question becomes:

Is (E) (DE) – in effect is the set (E) included in the set (DE)?

Simplifying further the question is reduced to:

Is (E)?

At which point the program recognizes the fact that 'is' followed by a set indicates that the correct answer is 'yes, E'. In the same way a command to 'Pick the large bolt to the left of the large nut' would be simplified through successive stages as follows:

Pick the (ABCF) (ADF) to the left of the (ABCF) (BCE).

Pick the (AF) to the left of (BC).

Pick the (AF) (A).

Pick the (A).

Although programs of this kind are obviously understanding natural language, albeit of highly simplified kind, it is far from evident that they can be said to understand it in our sense of the word. Suppose, for example, that it was necessary for the nut and bolt robot to add washers to the repertoire of objects that it could identify and manipulate. If, again, there were large washers as well as small ones, the machine would have to be programmed to distinguish between them as two quite separate objects and there is no way in which the knowledge of what constituted a small nut or a small bolt could be used to enable the machine to identify a small washer – in fact, the program has no understanding of the word small at all, it is only a symbol.

Although this example of a scene containing nuts and bolts of different sizes is a relatively elementary one, it makes explicit the principle which all language using programs rely upon to make an input in natural language intelligible to a machine. The Parsing technique is a method or algorithm for reducing any form of words, by the successive application of grammatical rules, until it has reached a point at which the computer 'recognizes it' as a logical formula to which it can respond. In order to perform the reverse process, that of creating a sentence in natural language out of one of the logical statements which the computer understands, the program will again need a data base which relates the symbols it manipulates internally to the words of its vocabulary, and a set of rules for converting one into the other. It is very relevant to the dispute referred to in Chapter Seven, between those who believe, with Chomsky, that man is a pre-programmed language user and the followers of Piaget, who hold that language is a learned skill, that many language using programs now employ rules that are based on Chomsky's own generative grammar. In theory at least, if such rules were complete and accurately formulated, they should, when embodied in a program, put a computer in exactly the same position, *vis-à-vis* natural language, as a human being. Unfortunately, though the programs allow computers to construct impeccably grammatical sentences they contribute nothing to its understanding of what those sentences mean – they are like the politician of whom it was remarked that 'he says nothing, and he says it very well'.

Perhaps the most telling attack that has been made on the top-down approach to language over this question of the level of understanding which is achieved was launched by the American philosopher John Searle in a paper entitled 'Minds, Brains and Programs'. The mode of artificial intelligence Searle singled out for criticism was that employed in the 'script using' programs developed in the United States by Roger Schank and R P Abelson. Though, in so far as they are concerned with the use of natural language in isolation from vision or other perceptive faculties, these models have little direct relevance to robots, they do serve to illuminate the problem of what is meant by 'understanding' in artificial intelligence in a particularly clear way. Briefly, one of the ideas behind the programs is that by providing rules very similar to the parsing program we have examined, computers can be enabled to 'map' one statement into another by rearranging the words contained in the original. This technique allows a computer to conduct conversations or answer questions about some simple story which has been supplied to it.

Suppose, for instance, that a program of this kind is supplied with the story of Jack and Jill, who went up the hill to fetch a pail of water. Given a grammar containing some relatively simple rules, such as that proper names start with capital letters, that nouns are preceded by the definite or indefinite article and linked by conjunctions to form noun phrases, it is perfectly feasible to design a program which will respond to questions such as 'What did Jack and Jill go to fetch?' or 'What did Jack break?' with answers like 'a pail of water' and 'his crown'. This is achieved by shuffling around the verbal components according to set formulas; and it would be beyond the scope of a program to answer the first question, as a human listener might, by replying '*some* water' or, in the case of the second, 'his *skull*', unless it is supplied with extraneous information, such as the fact that the purpose of the pail is simply to contain water, or that the word 'crown' is a synonym for the dome on the top of the human head which is formed by the skull bones.

In his paper, Searle postulated a group of people, none of whom could speak or read a word of Chinese, who were shut up in a room and given a short story written in Chinese characters, together with a complete and detailed set of rules for deriving answers to questions about the story from the written characters. None of these rules would take the form of translations of particular characters, but would be rules of thumb for rearranging them. Thus the answers the team gave would be in the form of characters taken from the script and manipulated according to the given rules. Now it is clear that even if they successfully carried out such an exercise none of the participants would understand the Chinese language any better at the end than they did when they started; it is also clear that their behaviour would have been similar to that of the script using programs. It follows that if we say that such a program 'understands' the statements it produces we mean something very different from the understanding that a human being has.

It could be said that Searle's group of people, or a script using program, understands a language only in the sense that an expert system may understand medicine: that is, it has a collection of data in the form of symbols and a set of rules for manipulating those symbols, but challenged to order a five course dinner in a Cantonese restaurant or to discuss the finer points of Mah Jongg with a Mandarin speaker, the individuals concerned would be as helpless as an expert system primed with the most advanced information about gynaecology which was confronted with a patient suffering from the common cold. The point is that although both language and medicine can be organized into structures of information which lend themselves to programming techniques, these structures by themselves do not result in an understanding of the underlying realities; for the abstractions were first achieved by a process of induction, deriving general principles from particular instances, and only become useful in the real world when one has some experience of the actual phenomena they explain. Human beings have very similar problems when they are asked to think about matters which are outside their experience; it is difficult, for example, to explain skiing to someone from Thailand who has never seen a flake of snow; and many of us boggle when faced with the notion of a sub-atomic world where there are infinitely many dimensions and particles which are sometimes particles and sometimes waves, whereas physicists, who have become familiar with the idea, find it relatively easy to live with.

Searle acknowledges that a program which could learn a language by a process of induction would be in a different category from one which, like his group of non-Chinese speakers, simply arranged symbols according to deductive principles. Such a program might be a learning automaton of the kind discussed in Chapter Seven, for the whole process of constructing a world automaton is an inductive one. And it could be assumed that such an automaton would be able to enlarge its vocabulary and diversify its usage of language as its world automaton (representing the words it 'knew', or could relate to its experience, and the ways in which they could legitimately be combined) grew. A learning process of this kind is difficult for any of the language using systems we have discussed simply because it is in the nature of a preprogrammed machine that its world automaton must be 'given' by the program and the extent to which it can generalize is severely limited.

The point about generalization is crucial, for the mastery of language hinges upon our ability to attach general meanings to words which, in practice, we encounter only in a limited number of specific instances and to adopt usages from one context to another. Thus, the ability to generalize would allow a machine, having learned the sentence, 'the cat sat on the mat' to formulate new ones such as 'cats sit on mats' or 'the dog sat on the mat', while the acquisition of experience, in the form of a sophisticated world automaton, would lead it to reject the possibility of 'the mat sat on the cat'.

The whole problem of enabling mechanical intelligence to learn by experience and to generalize is very closely related to the philosophical problem of the nature of 'universals'. A universal, in philosophy, is that general concept which, it is assumed, must be present in the mind before a word can be used in a particular context. Thus, the use of the phrase 'table manners' implies an agreed meaning of the word table which allows us to employ it in both a specific and a general way, so that we can say to a badly behaved child, 'Leave the table', meaning a particular table, but can also describe the same child as 'behaving badly at the table', meaning any table on which a meal is being served. The philosophical debate, which has continued since classical times, concerns the exact status of universals. Do they have an existence separate from and more real than that of particulars? Or is a universal simply the mean or average of all the particulars we have encountered? Artificial intelligence is clearly very deeply committed to the first point of view, for its prospects of success depend upon programmers being able to supply their machines with ready-made universals in the form of algorithms.

If, for example, an AI program is to allow a machine to discuss or describe the position of a pyramid, then the programmer has got to include in his instructions a full set of rules for identifying pyramids, or, in other words, he has to devise an algorithm for the universal pyramid. Winston's semanic nets offer some hopes of a way around the combinatorial problems of supplying a robot with a prefabricated universal for every object it will encounter, but, as we have seen, they quickly run up against the difficulty that a net which is loose enough to encompass all tables as seen from all points of view will inevitably accept as tables many objects which are, in fact, not tables; and one that is more tightly woven will reject many perfectly genuine tables. What is lacking in the machine is the additional element which human intelligence incorporates in the form of an idea of what tables are *for*.

It is this inability to perceive the meaning of words or objects in terms of their purpose which Searle argues is the fundamental weakness in the AI approach. The futility of the manipulations performed by his team of non-Chinese speakers was due to the fact that, however complete the set of rules that they followed, they lacked what he calls 'intentionality', that is to say they had no idea of the meaning or purpose of their manipulations except as an expression of the rules they had been given. One could argue that all AI programs are of this nature. The machines on which they are run are in the position of a human trainee who, having never seen an airplane and remaining unaware of the existence of such a thing, is taught to 'fly' on a ground simulator. Such a person may, in time, become a highly accomplished 'pilot', but if he is never allowed to know that there are such things as real airplanes the whole operation will be meaningless.

The script using programs of Schank and Abelson which were the immediate target for Searle's attack were in fact an attempt to deal with one symptom of this absence of intentionality, that which arises from the ambiguity of words which are used differently in different contexts – a difficulty closely related to the problems of computer translation at which we glanced earlier. A program like SHRDLU is of course script using, but it has only one script. Its vocabulary is limited to that which it requires to describe the scenes with which it deals – it needs to know about cubes and table tops and about picking and placing, but it does not need to know about spheres and table legs or about digging and throwing. Schank and Abelson suggested that if a program was equipped with several different scripts, each with a separate vocabulary, it would be able to 'talk about' several different subjects without becoming confused. Thus a script using program, given a 'movie going' script and a 'landscape' script, would know that, in a conversation where the words 'cinema' and 'Hollywood', which belong to the movie script, had already cropped up, the term 'blue film' would refer to pornography rather than colour, whereas if valleys and rivers had been discussed it would pick out the colour related meaning in the phrase 'blue sky'.

Schank and Abelson argue that the script using mechanism mirrors some underlying reality in the way human beings use language; we decide the meaning of words in particular contexts, they suggest, by referring to a list of scripts in our mind and selecting the one which applies. This line of thought was pursued by Mark Colby at Yale who developed programs which reproduced the behaviour of people exhibiting symptoms which caused them to be described as paranoid. Such behaviour is, he suggested, caused by a malfunction in the human script using mechanism that leads patients to stick rigidly to one particular script, even in inappropriate circumstances. Thus Colby's paranoid program, PARRY, relentlessly twists every conversation round to the subject of the Mafia which, it assumes, is pursuing it.

It must, however, be said that the script using device lacks conviction in the minds of many people for it seems a way of sidestepping the problem of meaning rather than confronting it head on. Schank and Abelson argue that given enough scripts of sufficient sophistication and the admittedly very extensive computing capacity to manipulate them, script using machines will produce behaviour that is convincingly 'human'. But to many this seems equivalent to suggesting that, because bears can be trained to ride bicycles, a bicycle riding bear is behaving in the same way as a human rider. Again, one returns to the problem of meaning and intentionality. Bears ride bicycles and, perhaps, script using programs manipulate their scripts as an ingenious trick which is achieved after much laborious human effort, but people ride bicycles, or use language, because they want to get somewhere or communicate something.

The fact that bears *can* be trained to ride bicycles does, however, lead to another question which is perhaps even more fundamental to artificial ingelligence for robots. Suppose, for a moment, that one wished to teach a computer to ride a bicycle. (This is not quite as improbable a goal as it sounds, for the problem could be tackled in principle as a model which need not involve placing an electronics kit on the saddle and equipping it with arms and legs; and it would be considerably easier to construct a model of a bicycle's behaviour than, say, the models of aircraft behaviour which are employed in flight simulators.) In order for the computer to balance the 'bicycle' it would have to be programmed with algorithms that instructed it which way to 'lean', how hard to 'pedal' and how to manipulate the 'handlebars', which is, of course, the key to maintaining the balance of a real bicycle.

But it seems improbable that even human children, let alone bears, actually learn to ride a bicycle by learning a complex set of algorithms. If one asked a six year old who had just achieved his first, wobbly ride across the lawn what he had learned, he would probably say that he had now got the trick or the knack of the thing.

Indeed, it is quite likely that he would be totally unaware of the fact that he remained upright because he had learned to keep his weight central and to compensate for a slight tilt to left or right by twisting the handlebars in order to move the base of the bicycle beneath its centre of gravity. This mechanism only becomes obvious when you try and ride a bicycle very slowly and have to turn the front wheel back and forth with increasing urgency to stay upright.

There is a very similar accomplishment which, in different forms, both humans and computers can master. Children, even performing animals, can learn to balance a stick upright on the end of a finger or a ball on the tip of the nose. This is clearly comparable to the job of adjusting the thrusters of a rocket during its launch phase in order to stop it toppling sideways. Yet the child, again, masters a 'knack' (one, perhaps, where the method used is a little more easily discernible than is the case with bicycle riding), whereas the need to keep a rocket at Cape Kennedy in the vertical position requires the attention of colossal computer banks in Houston constantly calculating and recalculating the complex series of equations that govern the vehicle's stability.

The point of this comparison is to suggest that when a child – or a bear – masters a physical skill, what is learned is a pattern rather than a set of algorithms. And this can be seen even more clearly if we return to the subject of vision. In so far as we ourselves understand how we interpret what we see, the idea that we do so entirely by testing every image against some store of algorithmic templates is simply not convincing – we could illustrate this in a hundred different ways, but one simple example may serve to clarify the point. All of us probably own a garment, a dress, a shirt or a coat, with a familiar and distinctive pattern which we would recognize instantly if we were to see someone else wearing a similar piece of clothing made of the same material – it is quite a common experience to misidentify a stranger as a friend or even a family member only to realize a moment later that the mistake is due to the fact that their clothes look familiar.

Now if the pattern on the garment were recognized because the brain had constructed an algorithm, something similar to a matrix or a semantic net, which defined the pattern, one would expect that it would be possible to recall the pattern in some detail; certainly to the extent, for example, of saying that a tweed had a houndstooth pattern rather than a checkered one or that the stripes on a skirt were made up of repeated sequences of red, blue and yellow. In fact, in most instances the best that the memory can summon up is a rather vague and indeterminate picture which 'goes out of focus' very quickly if one tries to zero in on details.

Yet the memory is obviously very distinct indeed at some level since we rarely fail to recognize such a pattern when we see it, or to distinguish it from relatively subtle variations. The distinctions between 'little black dresses' or blue suits with pinstripes, for instance, are sufficiently fine to defy precise description, but a person who has such a garment rarely meets someone wearing something they perceive as an exact replica.

Examples of this kind, and the reader will no doubt be able to think up many similar ones, suggest that the storage and recognition of patterns of one kind or another may be a more fundamental facet of intelligence than the manipulation of algorithms. This is not to suggest that the top-down model of intelligence which AI employs is valueless or unrelated to our own performance; there are many circumstances in which it is very clear that we use algorithms in our own thinking. In the case of vision, for instance, no process of 'pattern recognition' could plausibly eliminate the need for 'scene analysis' of some kind. But there does seem to be a strong case for suggesting that the logical, preprogrammed mode of thought is a refinement superimposed upon a more basic pattern recognition mechanism. It may even be that we share this latter technique with much of the animal kingdom and that our distinctive, 'human' brand of intelligence is due to our possession of an ability to handle algorithms or manipulate symbols which other creatures lack.

Following this line of argument one might come to the conclusion that the failure, or at least the disturbingly disappointing progress, of the top-down preprogrammed approach which AI has adopted might be due to the fact that it cannot realize its full potential unless it is underpinned by a more fundamental, though perhaps less formally 'intelligent', system operating as a pattern recognizer. Though this need is more evident in the case of vision it is by no means restricted to it. We saw in the case of the bicycle that it is possible to think of muscular coordination as a process of recapitulating patterns, and if we look back to the case of chess playing, it will be recollected that one of the differences between a machine and a skilful human player was that the latter seemed to have an ability to recognize the implications of certain states of the board, or patterns of pieces, without calculating their consequences by a logical application of algorithms which a decision tree would require.

Even in the use of language, pattern recognition can explain human accomplishments and experience in a way that preprogramming cannot. We are all familiar with the uncompleted sentence such as 'Well, I guess it's time to . . . ' or 'I think I'll just go and . . . ' which, depending on the context, is easily recognized as a signal that the speaker intends to go to bed, to have a bath, to cook dinner, or whatever. In such cases it is not only the form of words which follows a familiar pattern but also the circumstances in which they are spoken. The importance of pattern to language use can be seen in speed reading techniques, which rely upon the fact that, once mastered, the business of reading is a process of absorbing a pattern of words rather than resolving the meaning of sentences algorithmically as a parsing program would do. Speed reading courses train the reader to absorb larger and larger patterns at a single pass and it is difficult to reconcile their success with the sort of methods used by AI.

It is when we look for a means of achieving the recognition of patterns without resort to algorithms that the possibilities of alternative electronic architectures to that of the von Neumann computer become most interesting. The computer itself is disqualified as an effective pattern recognizer, though not for what might seem to be the obvious reason. Prima facie it might appear that its total dependence upon a given program would be fatal; after all, how can a machine that virtually depends upon the provision of algorithms in the shape of a program operate as an unprogrammed device? However, as we saw in Chapter Eight, when discussing the computer's ability to behave as a learning automaton, it is perfectly possible, given some basic programs to direct its operations, for a computer to behave like an unprogrammed, learning automaton. The difficulty with pattern recognition is slightly more subtle.

In terms of automata theory, any intelligent behaviour on the part of a machine (or a person) can, in principle, be thought of as the manifestation of an internal state structure. In the case of vision, what is required is a system which allows an automaton to change into a particular state in response to exposure to a given pattern, and to memorize that state and return to it when the same pattern is repeated. Preprogramming methods, of course, are intended to achieve exactly this result. When a machine that uses semantic nets compares the net produced by a particular table with its stored description of a 'universal' table, it will, if the two match, change to a 'table' state in which it will say, in effect, 'What I am seeing is a table.' In other words, learning a pattern requires that an automaton construct another pattern internally, in the shape of the 'if . . . then . . . ' statements which define a new state and that, when the pattern is repeated, it returns to that state.

If we take the case of another kind of change of state, that which occurs when a pattern is copied on to a piece of paper, there are two ways of bringing it about. The pattern can be drawn, freehand or by tracing it, in which case the 'state' of the paper will change as the work progresses and the task of ascertaining where each line should go is essentially an algorithmic one – the artist must refer to the original and, perhaps, measure the position and the length of each line. This process is analogous to the performance of a preprogrammed automaton, with the difference, of course, that the state of the paper is an external rather an internal one. But the alternative, printing the pattern on to the paper from a plate, produces a change of state that is instantaneous – one moment the paper is blank, the next moment the pattern is there. A psychologist might compare this to a change of *gestalt*, when a patient's whole perspective on, or feelings about, a problem are changed at one blow.

One way of approaching the task of pattern recognition is to seek a means of translating a pattern which is input to a machine into a change of internal state by such a single, *gestalt* process. But this is something that a computer is physically simply unable to achieve for it can, as we saw in Chapter Four, only alter its internal state bit by bit (or, more accurately, byte by byte). Where a change of state represents, or brings about, physical movement which takes place over a period of time and is, therefore, linear and sequential, this limitation does not much affect the computer's performance. But when intelligence must be applied to recognizing a visual image the handicap becomes very significant. For even simple patterns contain, as we shall see in the next chapter, immense amounts of information and even if the constituent bits are processed at a rate of tens of thousands every second, the computer cannot achieve the feat of recognition in 'real time', that is, as fast as images can be presented to it.

It seems clear, on the other hand, that the brain does undergo sudden 'catastrophic' state changes of this kind. This belief can be supported in two ways. First, in so far as we can keep track of them, our own processes of thought and perception do not always seem to take place in a gradual linear way: it is, to give but a single example, a fairly common experience to find that the solution to a problem or the grasp of an idea which has consistently eluded one, appears as if by magic: one moment all is murky and confused, the next it is clear and obvious. Second, what we know of the structure of the brain suggests very strongly that it is a mechanism which one would expect to change states suddenly and in great leaps rather than short, single steps. The way that the neurons within the brain are interlinked, by complex webs of nerves which transmit pulses from one cell to another and feed them back again, indicates that a thought is more akin to a tidal wave sweeping through the structure than to the kind of orderly flow which occurs within a computer.

Because a computer is a universal machine, it is naturally possible to program it to behave like a network of neurons, but though the result may be interesting and informative, it will not be useful in the real world which robots inhabit, for the task of simulating the kind of change that takes place in a neural network instantaneously may well preoccupy a computer for a period of minutes or even hours. If we want to understand how the brain works *and* make practical use of the knowledge, the only answer is to try and build machines that are, as far as their characteristics as information processors go, actually like a brain.

The remainder of this book is, therefore, devoted mainly to machines which can achieve this trick of altering their state instantaneously and 'catastrophically'. We also hope to show that machines of this kind, built from the bottom-up by following Wiener's proposal for artificial neural nets, can go a long way towards supplying the capacity to learn from experience, the ability to generalize by induction, and the intentionality which we identified as the major problems left unsolved by the top-down approach to artificial intelligence.

10. The Silicon Neuron

All nature is but art unknown to thee;
All chance direction which thou canst not see;
All discord harmony not understood;
All partial evil, universal good.

Alexander Pope, 'Essay on Man'

The main reason why, for a quarter of a century after the publication of *Cybernetics*, Wiener's proposal that mechanical intelligence might be investigated on 'bottom-up' principles went unheeded was the apparent impracticality of the undertaking. However intriguing the idea of direct comparisons between natural and artificial neural nets, it simply did not look like a workable proposition on either technical or economic grounds.

The overriding problem was stated very clearly by the parent of the preprogrammed machine, John von Neumann himself, in a series of lectures delivered at Yale University in 1956. Von Neumann was mainly concerned with a slightly different point, for he was anxious to emphasize the fact that disparities in scale made any comparison between contemporary computers and the brain premature to the point of absurdity. But the argument he advanced applied with even greater force to 'alternative architectures'. The largest computers then in use had a storage capacity of some 10^6 bits, that is to say their electronic memories contained some million switches, each storing one bit of information. In contrast, the best current estimate is that the number of neurons in the brain is approximately 10^{10}, or ten thousand million, and that each neuron may have a capacity measurable in tens of bits, or the ability to store several thousand different messages. Using existing technology, the construction of a computer with a comparable memory capacity would have absorbed a sum of money that was mind boggling: at the price levels then ruling, it would have sufficed to purchase ten thousand million family cars.

The whole idea of building a neural net of any size was pushed even further into the realms of fantasy by the fact that the elements would have to be interconnected to form a network rather than having all their input and output circuits channelled through a control unit and central processor as in an orthodox computer system. What von Neumann could not foresee, and due to his tragically premature death in 1958 was never to know, was that the phenomenal success of his own brainchild would engender a revolution which altered the whole basis of his calculations. With the arrival of the first silicon chips in the mid-1960s, the economic factors began to be less and less significant; indeed, if the cost and size of solid-state electronic components continue to diminish at the rate which has been achieved over the past three decades, it will be possible in the year 2000 to purchase a brain-sized collection of chips for the cost of a single car and to fit them into something the size of a shoe box. Already manufacturers are experimenting with single chips which store 256,000 bits, a capacity comparable to that of quite an appreciable fraction of the brain.

The first questions that arise in any project for an artificial neural net concern the kind of circuits required and the 'architecture' that should dictate their assembly. The obvious starting point for a bottom-up design must be a detailed examination of the neuron that is the basic information processing element in the brain itself, followed by an attempt to find some artificial counterpart to it. It is not, of course, necessary to try to mimic the actual physical structure or the chemical constituents of a nerve cell; what is significant is the way in which it deals with the digital pulses of electricity that are the brain's equivalents of the computer's binary digits.

Thus far we have contented ourselves with describing neurons as the natural equivalent of the electronic switches that store the binary 1s and 0s in a computer's memory. Though accurate enough as far as it goes, this does not tell us very much about a single cell's function as a processor of information, as opposed to a mere repository for it. In particular, it does not reveal anything about the circumstances in which a cell accepts a digit into storage and transmits one by 'firing', or generating an electrical pulse. If we look at the knowledge which neurophysiology has managed to glean about the actual behaviour of individual neurons, it becomes clear that the operations of the cell are a good deal more subtle and complex than those of a simple switch.

By the use of electrical probes, which have now been reduced to a size which makes it possible to measure the electrical output of a single neuron, scientists are able to determine when a cell is firing.

High powered microscopes also allow us to trace the interconnections between neurons. Each cell can be seen to have one noticeably dominant nerve leading out of it; this is its 'axon', and when the neuron fires, a burst of pulses travels out of the cell along the axon. For all practical purposes, the burst can be thought of as a single pulse which, physiologists agree, is equivalent to one bit of information. Some axons divide up into a multiplicity of other nerves, each leading to another neuron (though each tributary nerve will carry the same message at the same time in the form of the presence or absence of a pulse), while the pulses from other axons travel direct to the muscles along the efferent nerves.

It is also known that what causes a neuron to fire is the pattern of pulses it receives from other neurons. Each cell may simultaneously receive pulses from the axons of thousands of others, and the way in which it interprets their message is not straightforward. To understand what happens within the neuron to determine whether it fires, it is helpful to think of the decision making process as taking place in two stages. In the first stage, the neuron can be thought of as a ballot box to which each of its neural neighbours contributes either a vote (in the form of a pulse) or an abstention (no pulse). The neuron, however, does not act in response to a single vote or even accept a majority decision, for each pulse is received at a point on the periphery of the cell called a synapse which, in effect, then has the power to 'rig' the ballot. At this second stage, when a synapse receives a pulse, it determines what that vote means, or how much weight it should carry in reaching the decision to fire or not. A synapse can, for example, attach a value of 0, +1 or -1, or even +2 or -2, to an incoming pulse, and it is these values which, in combination, decide whether the neuron fires.

To clarify the idea, consider a simplified picture of a neuron which is connected to only three other neurons, A, B and C, and which fires when the 'votes' in favour total 2. Suppose, first of all, that the synapses which are connected to A and B give a value of +1 to a pulse and that connected to C counts a pulse as -1. The behaviour of the neuron can then be described by a table as shown overleaf (top).

In this instance it can be seen that, out of the eight possible combinations of pulse and no-pulse, only one will cause the neuron to fire. But, if the synapses changed their properties, so that a pulse from A scored -1, one from B, +2, and one from C, 0, and the neuron con-

Pulses Received (*)			Value Determined by Synapse			Result
A	B	C	A	B	C	
–	–	–	0	0	0	0
–	–	*	0	0	–1	–1
–	*	–	0	+1	0	1
–	*	*	0	+1	–1	0
*	–	–	+1	0	0	1
*	–	*	+1	0	–1	0
*	*	–	+1	+1	0	2 (fire!)
*	*	*	+1	+1	–1	1

Pulses Received (*)			Value Determined by Synapse			Result
A	B	C	A	B	C	
–	–	–	0	0	0	0
–	–	*	0	0	0	0
–	*	–	0	+2	0	2 (fire!)
–	*	*	0	+2	0	2 (fire!)
*	–	–	–1	0	0	–1
*	–	*	–1	0	0	–1
*	*	–	–1	+2	0	+1
*	*	*	–1	+2	0	+1

tinued to require two votes to fire, the table would show that firing occurred in two out of eight eventualities (lower table).

It has been inferred that such changes in 'synaptic weight', causing changes in the functions of a neuron, do occur, and that the highest rate of change occurs when the owner of the brain is young. It should not, however, be concluded that the loss of flexibility as the number of neurons susceptible to change decreases means a loss of mental powers; rather, the increase in the proportion of neurons whose functions are fixed and 'frozen' represents the increase in the amount of information stored in the memory.

A neuron with synapses that changed the weights they attached to pulses arbitrarily or at random would obviously serve no purpose; in order for the system to 'mean' anything, changes must come about in some organized way.

If the neuron's property of firing only under certain conditions can be thought of as a capacity to remember its function, then the ability to modify those conditions represents a capacity to 'learn' something new. In fact, the neuron we have described can be thought of as a miniscule automaton, containing the instruction, 'If such and such a pattern (or patterns) of pulses is received then fire,' and a change in synaptic weights is equivalent to a change in the neuron's state. A change in synaptic weights must obviously only take place when the system as a whole 'decides' to learn a fresh function. It is possible to imagine that some mechanisms similar to those employed by the learning automaton described in Chapter Seven are at work here: for if one thinks of the section of the brain which is 'learning' as one automaton, then it is likely that the decision that it should learn at a particular moment is taken by a secondary automaton.

For present purposes, however, it is enough to know that changes in synaptic weight are a result of the fact that certain synapses are 'dominant'. When a pulse is received through a dominant synapse it has the effect, first, of making the neuron fire whatever its current function might dictate, and, second, of causing changes in the weights of the other synapses so that whatever pattern of pulses is present at that moment will, in the future, constitute a 'firing' pattern. In effect, the dominant synapse is rather like an army instructor who displays a silhouette of an enemy plane and tells his pupils, 'Whenever you see that shape again, shoot.' If we consider the neuron as an automaton, then the dominant synapse has a role similar to that of the teach button on the learning robot in Chapter Seven, for it tells the neuron when it is required to learn and remember a new state.

Despite the difficulties identified by von Neumann, several attempts were made during the 1960s to model the behaviour of comparatively modest networks of neurons. Two of the most notable of these, the 'Adaline' device developed by Bernard

Widrow in the United States and a machine constructed at University College, London by W K Taylor, used elaborate machinery to simulate alterations in synaptic weights. In the case of Adaline, the firing behaviour of the 'neurons' was changed by electro-chemical means; Taylor's machine achieved the same effect by the use of variable resistors, similar in principle to light dimmers, powered by electric motors. Both projects suffered from the problem that they had to employ roomfuls of elaborate equipment to simulate even small networks of neurons, and it was obvious that such machines could never, realistically, be extended to a size at which they would be useful let alone aspire to the dimensions of a brain; an additional handicap in the case of Adaline was that the electro-chemical system was slow and highly unreliable in practice.

The only possible alternative to using contraptions like these seemed to be to model the behaviour of a net by use of computer programs. This was the basis for the 'Perceptron', developed by Frank Rosenblatt at Cornell University towards the end of the 1960s. Though the computer's universality makes it quite possible to program one to simulate the behaviour of a network of neurons, the operation is lengthy and laborious. In the neural net, where each cell fires according to the rules we have described and feeds its pulses to scores of other neurons, causing them to feed back further pulses that in turn result in other neurons firing, the consequences of feeding just a single pulse into the system can be dramatic and totally unforeseeable; moreover, the process of changing the state of the system as a whole takes place, to all intents and purposes, at a stroke. To calculate the behaviour of a net by sequential computation of the role of each individual neuron is rather like trying to assess the effect of a cloudburst by calculating the action of each separate drop of rain. The practical problems of modelling or simulating neural nets were compounded by a most eloquent critique of Rosenblatt's methods produced by Marvin Minsky and Seymour Papert of MIT in 1969. This effectively put an end to work along these lines.

But, happily, and entirely fortuitously, the silicon chip revolution in the latter half of the 1960s produced a device which turned out to be an ideal artificial neuron. Although the pioneers of silicon technology were mainly interested in reducing the size and easing the manufacture of computer logic circuits, the technology had advanced sufficiently by 1969 for attention to turn to the design of memory chips which would, in large part, replace the cumbersome magnetic store back-up memories on which all computers of any size depended for most of their capacity. The result was the first RAM (Random Access Memory) chips which are today a familiar part of every home computer.

The purpose of the RAM, in the context of a computer, is to provide the memory elements which store digital information, either data or programs, and it is normally divided into sections, each of which stores eight or more bits. In order to store, say, eight bits, eight electronic switches are required; in point of fact each switch is a device called a bistable made up of two transistors and having two possible states, representing either a 0 or a 1. Digits are placed into storage in the bistables through a 'data-in' terminal and read off, when required by the program, via an 'output' terminal.

For the memory to serve any useful purpose, it must naturally only accept information into a particular part of its store, or output the information held in that location, when instructed to do so, and this control is exercised by means of the ' address' terminals. An eight-bit RAM is so constructed that one of each of the eight possible combinations of three digits, from 000 to 111, at its three address terminals 'addresses' one of the eight bistables. If, at the same time, a 0 or a 1 is present at the data-in terminal, the bistable will be 'set' to store that digit, but if a bistable is addressed without any data being fed into it, the result will be that it outputs the digit it contains (though without deleting it from store).

We now take the case of two eight-bit RAMs which have been set in different ways. In the first, all the bistables have been set to store a 0 except the one with the address 101. In the second RAM, six bistables have been set to 0 and two, those with the addresses 010 and 011, to 1. We can now draw up a table (see overleaf) showing the output that will occur when each of the bistables in the two RAMs is addressed.

If this table is compared with those on page 212 showing the firing performance of neurons with different synaptic weights, it will be seen that the input of 1s or 0s at the address terminals of the RAMs corresponds with the presence or absence of a pulse at the neural synapses, and that the output of a 0 by a RAM corresponds with a failure to fire and that of a 1 with a firing. It can also be seen that inserting a fresh digit via the 'data-in' terminal, while addressing the

Address	*RAM 1*	*RAM 2*
000	0	0
001	0	0
010	0	1
011	0	1
100	0	0
101	1	0
110	0	0
111	0	0

bistable which one wishes to set, is equivalent to altering the synaptic weight in the neuron by delivering a pulse to the dominant synapse when a given pattern of pulses is present at the other synapses. Thus, by quite different means, the neuron and the RAM achieve very similar firing characteristics. From this it follows that a network of RAMs interconnected in the same way as a network of neurons will process information in a similar fashion.

Though it is tempting to press ahead at this point and examine the possibilities opened up by the prospect of constructing artificial nets of interconnected neurons, to do so would be to ignore the impressive properties of the relatively straightforward assemblies of silicon neurons known as 'single layer nets'. In a single layer net the RAMs are not connected to each other; instead each RAM receives an input of information separately and the output of the net is measured by the number of RAMs that are 'firing'. The context in which the potential of single layer nets has been most fully explored is that of visual pattern recognition systems, so it is on this application that we concentrate here.

To make the mechanism absolutely clear, we shall start by considering the very simple patterns which can be placed on a row of three squares. This is equivalent to a highly simplified television image in which there are only three pixels, or picture points, each of which must be either wholly black or wholly white. There are eight possible arrangements of black and white squares:

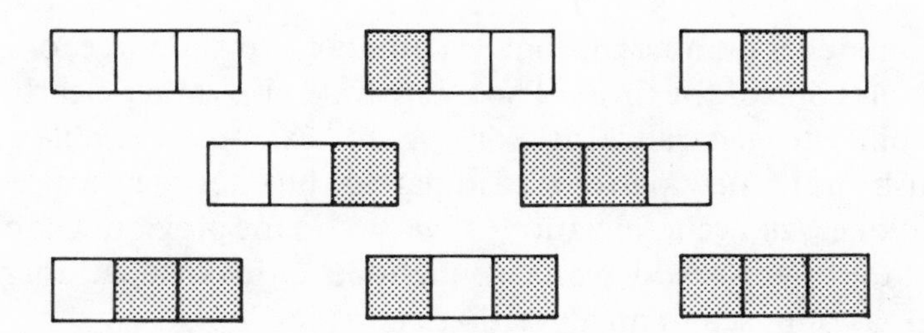

Patterns formed by a binary three pixel image

If each of the squares is connected to one of the three address terminals of an eight-bit RAM of the kind described above, with the left hand square providing the first digit of the address, and a black square translating into a 1 and a white square into a 0, then each of the eight addresses in the RAM can be used to represent one of the patterns. If, for example, one wished to teach the RAM to remember and recognize the pattern at the top right, that in which the centre square is black and the other two white, all that would be necessary would be to activate the address terminals, which would carry the pattern 010, while at the same time activating the RAM's data-in terminal. The result would be that the RAM would now be set to output a 1, or fire, whenever the same pattern was present at the address terminals. It would, of course, be possible to repeat the process with any or all of the other patterns so that the RAM fired when it 'saw' any of them; equally, by switching the system off and starting again, it would be possible to reset the RAM to respond to a different pattern, or set of patterns.

The simplest kind of net would obviously be one like this, consisting of only a single RAM. And it is possible to build very large systems of this kind, for RAM elements are not limited to the simple eight-bit capacity we have used as an example. It is quite possible to have RAMs with a storage capacity of sixteen bits, and therefore four address terminals and sixteen possible firing patterns; a thirty-two bit RAM will have five address terminals and thirty-two firing patterns, and so forth right up to RAMs with a capacity of many thousands of bits. The most obvious deficiency of a single, giant 'neuron' is that it has only two possible responses to any input – either it fires, outputting a 1, or it does not. In effect, it can answer only 'yes' or 'no' to any question that is put to it; so, even if it has been taught on three different patterns, it cannot say that a new image matches number one but not number two and three, only that it matches one of the three. The single RAM also rather clearly lacks any ability to generalize. It will not, for example, be able to distinguish that a new pattern is similar to, but not identical with, a taught one. And generalization, as we saw in the previous chapter, is one of the qualities which one would hope to find in a pattern recognition, as opposed to an algorithmic, system.

A way around the limitations imposed by a simple yes or no response would seem to be the use of a number of separate RAMs. How should they be deployed in order to provide the greatest possible capacity and versatility? One apparently elegant solution would be to teach each one a different range of patterns. If, for example, there were three RAMs, each of which had been taught four patterns out of a total of seven, so that the first RAM recognized, say, patterns 4, 5, 6 and 7, and the second RAM patterns 2, 3, 6 and 7, and the third RAM patterns 1, 3, 5 and 7, then it would be possible to identify which pattern was being recognized by treating the output of the RAMs as a binary code (in the example we have chosen pattern 1 would be coded 001, pattern 2, 010, etc). But though this expedient would enlarge the system's repertoire and enable it to identify one pattern out of a whole variety which had been taught, it still provides no generalization. The arrangement also has the disadvantage that, if utilized on a large scale, its output takes the form of yet another complex pattern of binary digits – an unfortunate trait in a machine that is supposed to *recognize* patterns not to *generate* them. Finally, it does not accord with what we know of the brain's own organization, an important consideration if we are adopting bottom-up criteria.

Our knowledge of the brain's methods of dealing with this situation is due to the work of two Nobel Prize winners, D H Hubel and T N Wiesel, who established (by laborious experiments with cats and monkeys) that different parts of a natural neural net respond to different kinds of patterns. If, for instance, the eye perceives an image which consists of a vertical bar, then firing activity occurs in a particular locality in the brain, but if the pattern is changed to a horizontal bar then a quite different group of neurons respond. As a consequence, when a vertical bar causes firing at point A and a horizontal bar firing at point B, the brain is able to recognize (or, perhaps, even 'understand') that what is being seen is a cross. But recognition does not seem to depend upon information about where firing is taking place being encoded; instead, the brain associates activity in a particular area directly with a specific kind of image. The difference can be illustrated by an analogy. The RAMs in the example above are like a coding machine which, when a key is pressed, creates a pattern of dots and dashes to represent a letter. The brain, in contrast, is more like a typewriter which responds to the depression of a key by printing the corresponding character.

If the engineer pays heed to these clues to the workings of the natural system he will obviously be well advised to try to organize his RAMs in a similar fashion. He might perhaps teach one to recognize vertical bars, a second to respond to horizontal bars and allot others to patterns involving diagonal bars. But although this expedient brings us closer to the natural model, it still holds out no prospect of generalization. Each of the RAMs will respond only to the exact pattern which it has been taught to recognize.

To see how generalization, and a response which is more sophisticated than a simple 'yes' or 'no', can both be achieved we will imagine that we have been asked to design a system that can distinguish between two simple patterns which occur on an 8×8 matrix, containing 64 pixels, each of which still has only two possible degrees of intensity, white or black. To take a case drawn from actual practice, we will suppose that one image represents the wheel of a toy car and the other a door. If a system of this kind could be made to work it could form part of a robot which sorted the parts into separate categories. For simplicity's sake we also assume that both images appear in a predetermined position and orientation so that they always occupy the same spaces on the screen. The two patterns might look something like those shown overleaf.

The simplest assumption would be that this task requires two RAMs, one taught to recognize the wheel and one to recognize the door. It is naturally important to establish how large the RAMs are going to have to be and this can be done by following the basic formula that determines the number of different messages that can be carried by a binary number of given length. The formula states that for 'n' number of digits the maximum number of possible messages will be two raised to the power of 'n', or 2^n. Thus two binary digits can convey any one of four (2^2) messages (00, 01, 10

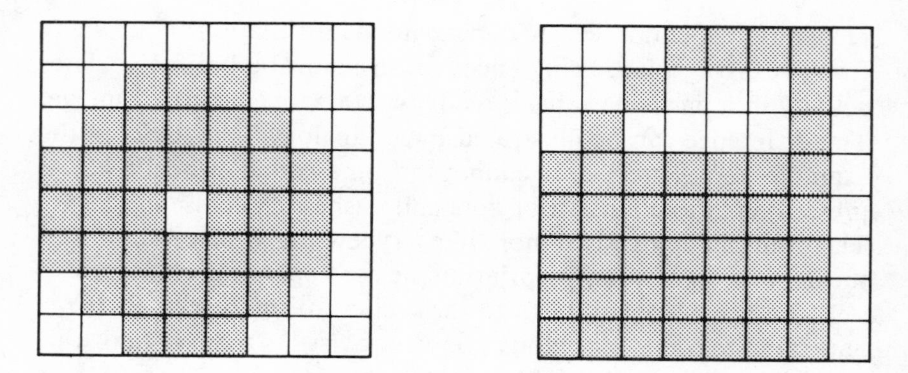

The wheel image (left) and the door image (right) on a 64 pixel matrix

and 11) and three digits can convey eight (2^3) messages (000, 001, 010, 011, 100, 101, 110 and 111).

Since each of the 8 × 8 matrices which contain the images has 64 pixels, each of which generates one digit of information (a 0 if the pixel is white and a 1 if the pixel is black) each RAM will require 64 address terminals and the capacity to store 2^{64} bits of information. This is a spectacularly large amount of storage, especially for a system which is to carry out such an elementary task, for 2^{64} is about ten million million million. It also remains the case that, out of this huge number of possible patterns, only one will produce a firing response from our giant neuron.

This raises a very practical issue, for it is a fact that, even if one was satisfied in principle with a machine which was so demanding in terms of storage and was still incapable of any degree of generalization, it simply would not work. It is inevitable that, in the process of translating the pixels into digits, some 'noise' will occur and cause distortions, turning 1s into 0s or vice versa. If we supposed that 10 of the 64 bits of information which defined the image of wheel were distorted by noise, which would admittedly represent an improbably high level of disturbance, the wheel picture might perhaps turn out rather like the image shown opposite. This is almost unrecognizable even to the human eye: to the RAMs, of course, even one distorted digit would be fatal.

It might be thought that the problem could be surmounted by training the RAMs on a whole range of typical distortions, so that instead of firing only in response to one of the possible patterns,

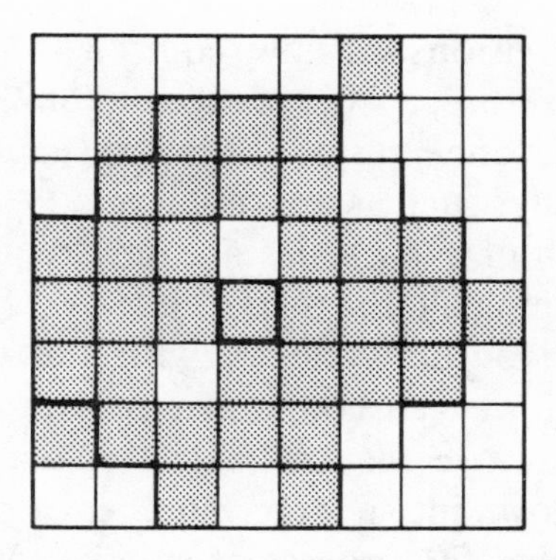

The wheel image with 10 of the 64 pixels distorted

each RAM also fired when it was presented with the patterns, which are close to, but not identical with, the perfect image. This would involve teaching each RAM the original undistorted pattern and then, as a separate operation, teaching it every possible pattern that might come about if noise caused a given level of distortion. If it were safe to assume that no more than one in 64 digits would ever be distorted at any one time, then 65 separate teaching processes would be needed. But the exercise quickly becomes impossible if more serious levels of distortion have to be provided for. In order to guard against the possibility that four out of 64 pixels might be distorted, for instance, it would be necessary to go through 679,140 separate teaching sequences. Even the most ardent advocate of automation could hardly recommend this undertaking to a toy manufacturer!

It must also be borne in mind that a system which is to be useful in real life applications would have to deal with images of a far higher definition than is provided by a 64 pixel matrix. A typical robot application would involve the use of an image made up of 512 × 512, or over a quarter of a million pixels and, therefore, an astronomically large number of possible distortions.

The way that has been found of circumventing this apparently horrendous problem has several advantages which are not obvious until one examines its implications in some detail. Basically, the trick is to use more and smaller RAMs. Instead of using one giant RAM to recognize each pattern, smaller units are put together to form what are known as 'discriminators', with each RAM being allotted one section of the image. Suppose that we now divide the 64 pixel image into four quadrants, each containing only 16 pixels.

Each of these subdivisions will now require a RAM with only 16 input terminals and, therefore, a storage capacity of 2^{16}, or about 256,000 bits, which is about one quarter of the RAM capacity of a medium-sized home computer such as an APPLE or a PET. Even if we allow for the fact that four such RAMs are now required to store each image, making eight in all, the total of just over two million bits still compares very favourably indeed with the enormously larger number with which we were confronted previously.

However, the storage requirements would quickly grow if we broke the image down into a larger number of pixels. So we might well decide to repeat the process of subdivision and increase the number of RAMs in each discriminator to 16, each covering only four pixels. The number of bits each RAM now has to store is reduced to 2^4 or 16 and the total for each discriminator is therefore 16×16, or 256 in all, an absurdly trivial capacity in modern computer terms.

The obvious question is, how has this dramatic reduction in the amount of hardware required affected the performance of the system? The answer is a rather surprising one. To illustrate the point, we will return to the intermediate version of the machine, the one that subdivided the 64 pixels into four quadrants of 16 pixels each and utilized four separate RAMs per discriminator. We will suppose now that this machine has learned to recognize not only the 'perfect' wheel image but also the 64 variations of it that could result from one pixel becoming distorted. Each of the RAMs will, as a consequence, produce a firing response when shown one of 17 patterns – the perfect version of its quadrant of the image plus the 16 distortions produced by a misreading of one of the 16 pixels in that quadrant.

The first point to note is that the response given by the machine as a whole is no longer a simple 'yes' or 'no', but instead a kind of ballot to which each RAM contributes one vote. Any of the 65 patterns on which each discriminator has been trained will obtain a unanimous, four out of four, response; but there is now a high probability that even a more seriously distorted image will obtain a majority vote in its favour. Imagine for a moment that the image of the wheel has been distorted by six picture points, one in each of three quadrants plus three more in the lower right hand quadrant, with the result that this quadrant now matches the door image exactly instead of the wheel one.

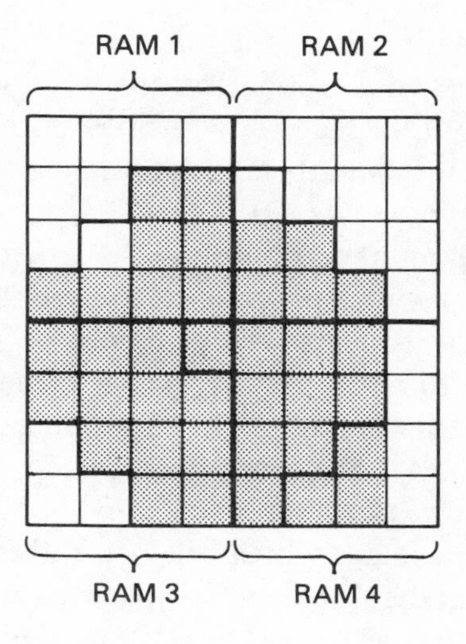

The wheel image distorted by six picture points, one each in the quadrants assigned to RAMs 1, 2 and 3 and three in the quadrant for RAM 4, which is now identical with the door image

In response to this image three RAMs out of four in the wheel discriminator will fire, and so will one of four RAMs in the door discriminator. In effect, the machine will say, 'this image is three quarters like a wheel and one quarter like a door.' Though this method of making decisions, known as 'maximum response', will obviously produce fairly crude answers when the image consists of only 64 pixels and is processed by discriminators containing only four RAMs, it is not difficult to imagine that, with an image resolved into half a million pixels and discriminators which contain dozens of RAMs, it is a highly effective way of arriving at conclusions which will be accurate over 99 per cent of the time even if considerable levels of noise are present. For the chances of noise being so prevalent and evenly distributed as to generate false responses in more than half the RAMs are very low indeed.

But an even more interesting consequence is that a system of this kind has acquired an extremely impressive ability to generalize. One way of expressing a capacity for generalization would be to measure the extent to which a system recognizes patterns to which it has not been exposed in training; in a sense, an ability to do this amounts to a capacity to create universals. For the point about our own ability to classify, say, the pattern of a jacket as a plaid is that we can do so even if we have never seen the actual pattern before. The act of recognition amounts to a comparison between the pattern we see and some kind of 'universal plaid' stored in our memory which allows us to say that this pattern is definitely a plaid rather than a check or a stripe.

To calculate the number of 'untaught' patterns to which the machine will respond with a unanimous four out of four vote, we go back to the training process which, it will be recalled, involved teaching each RAM to recognize the 'perfect' image in its quadrant plus the 16 other patterns that might be produced by the distortion of one pixel. Each RAM therefore fires in response to 17 possible patterns and it follows that there are 17 × 17 × 17 × 17 or 83,521 patterns of the whole image which will cause all four discriminators to fire in unison. But the system as a whole was only taught to recognize 65 patterns, the perfect 64 pixel image plus the 64 one-bit distortions of it. Subdividing the discriminators has therefore given each of them the ability to recognize with total certainty 83,456 patterns which they have never seen before.

Admittedly, there are a total of 679,140 patterns that could result from no more than four distortions in the image as a whole and the remainder of these will cause at least one RAM to fail to fire, since they only come about when one quadrant of the image remains perfect and another contains two distortions; but if we accept a three to one vote as confirmation that an image has been recognized, many of these additional patterns will be accounted for. As a matter of interest, it can be calculated that the system will respond with either a four to four or a three to four vote in favour of 84 per cent, or 569,840 out of the 679,140 possible patterns that can be produced by the distortion of four pixels. As we have seen, the machine will also give a three to four verdict to the many other patterns that result from a higher level of distortion concentrated in one quadrant alone.

Given the success that has been achieved by subdividing the image and increasing the number of RAMs, it might be thought advantageous to continue the process. The snags that arise from such over optimism become evident if we carry the idea to its logical conclusion. The smallest possible RAM is one with a single address terminal and the capacity to store only two bits. It would take 64 such two-bit RAMs to cover the 64 pixels of the images we have been discussing, so the total storage capacity of each discriminator would be a highly economical 128 bits. Trained on an ideal, undistorted image, the wheel discriminator would respond with 64 firings out of 64 to the wheel and with 50 out of 64 to the door (50 out of 64 picture points are identical in both images). The problem is that, if the system was then trained on the 64 possible one-bit distortions, both discriminators would give a 64 to 64 response to *any* image – what has happened is that every bistable in every RAM is now storing a 1. This stage of affairs is known as 'saturation', and one of the tasks of an engineer is to find the right compromise between a small number of RAMs and, consequently, a large storage requirement and a low level of generalization, and a large number of RAMs, with a small storage requirement but a tendency towards saturation.

It can be shown that the right answer for the 64 pixel images we have examined is 16 RAMs per discriminator, each containing 16 bits of storage. After training of the original patterns plus the 64 one-bit distortions, the wheel discriminator will now respond to the wheel with 16 out of 16 firings while the same image will produce only 10 out of 16 firings in the door discriminator. This is not such a high degree of certainty as that which the four larger RAMs produced, but it would be quite adequate in practice and generalization is almost complete with all images containing up to four distortions.

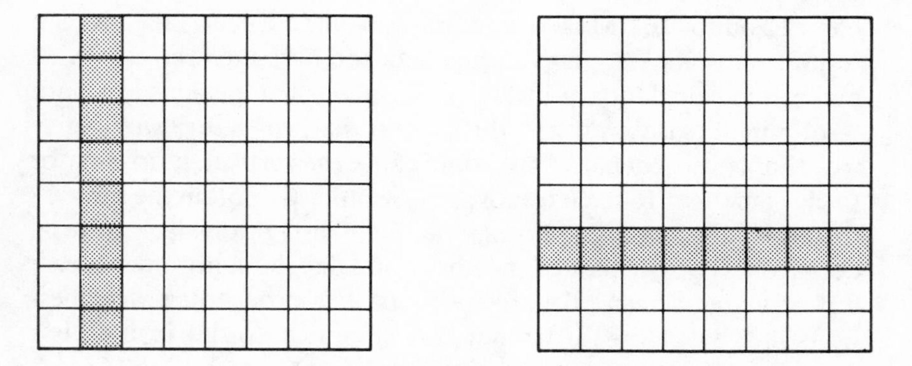

One-pixel bars on a 64 pixel matrix (see overleaf)

Though the effort to guard against the effect of noise has produced an impressive level of generalization, it is not the only factor which improves generalization in a single layer net of this kind. Consider now a different kind of pattern, that formed of a single bar, one pixel wide, running either down or across the 64 pixel image (see previous page).

There are clearly only eight vertical and eight horizontal bars on which it would be necessary to train the system. Subsequently, if separate discriminators have been assigned to each kind of bar, the system as a whole would have some general notion of 'vertical-ness' and 'horizontal-ness' which could be used to distinguish these properties in a whole variety of other images. If, for example, the discriminators had each consisted of 16 16-bit RAMs it can be calculated what response each would produce to an image like this:

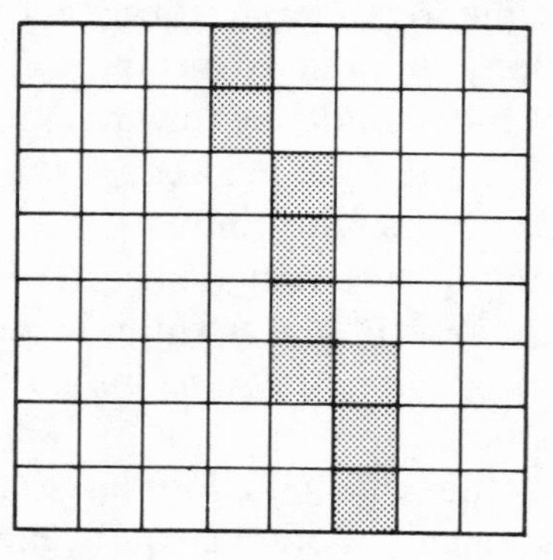

The RAMs that had been trained on vertical bars would give a 15 out of 16 response and those which had been trained on horizontal bars, 12 out of 16. (Most of the firings would, in fact, take place in response to blank squares rather than the bars themselves, which makes the calculation a little complicated.) The response of both discriminators is very nearly the same as that which they would give to a simple vertical bar. This would cause the vertical discriminator to give a perfect 16 to 16 response but would still obtain the same 12 to 16 response from the horizontal one. Though this lack of total decisiveness on the part of the horizontal discriminator might seem unfortunate, we can take advantage of it in order to answer questions such as: 'How far from a perfect vertical (or horizontal) is a given line?' Since the horizontal discriminator gives a 12 to 16 score to a vertical bar (and the vertical discriminator will do the same for a horizontal bar) we can simplify things by deducting 12 from both 'scores'. We can then say that the response of the two discriminators to a perfect vertical or to a perfect horizontal is four to zero and that, in the case of the mostly-vertical-but-a-bit-horizontal pattern above, the 15 to 12 score, which looked a little like indecision, becomes three to zero. With this new scoring system, a pattern like the one below, which is very nearly exactly intermediate between the vertical and the horizontal, will produce a response of one to minus one (11 from the vertical discriminator and 13 from the horizontal one).

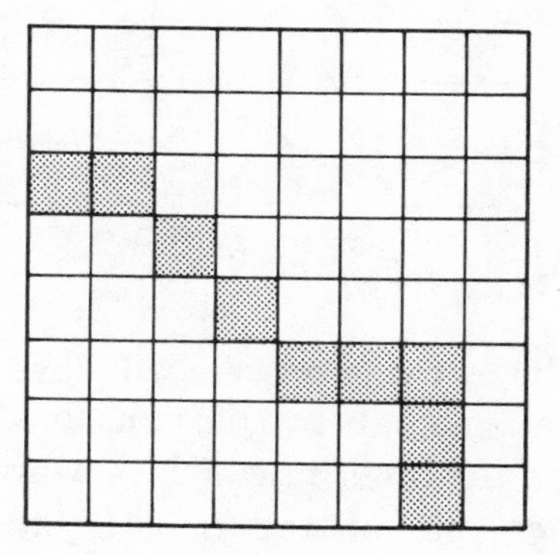

In this instance the system is saying, in effect, 'I am not very sure either way (a fact reflected by the low scores of both discriminators) but if I had to bet on it I would say it was horizontal rather than vertical.' If this system was shown the wheel image that we used earlier in the chapter, it would give a score of minus eight to minus eight, amounting to a conclusion that: 'This is absolutely nothing like a vertical bar or a horizontal bar.'

This property which enables the combination of two very simple machines (in terms of contemporary electronics) to detect the overall orientation of an image is attractively reminiscent, to the cybernetician, of the mechanism which Hubel and Wiesel discovered in the eyes of animals. No one would suppose that it was a substitute for a sophisticated, preprogrammed scene analysis device, but it might very plausibly be supposed to be a prerequisite for the success of one.

Now for a problem with a highly intriguing solution. Suppose that two 16-RAM discriminators are trained on, respectively, vertical and horizontal bars each *two* pixels thick, as shown overleaf.

The training process would consist of showing each discriminator four non-overlapping bars. But, remembering that each separate RAM is 'looking at' only a single square of four pixels, it is evident

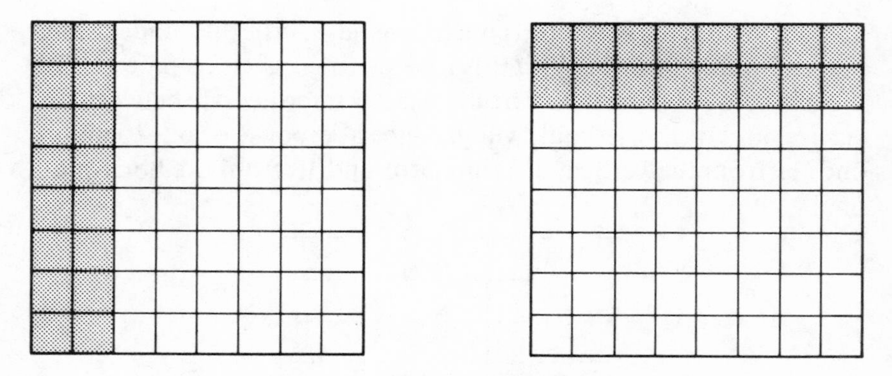

Two pixel bars

that the result of the teaching process will have been to show each RAM in the discriminators only two patterns, one in which all its four pixels are white and the other in which they are all black. In consequence, whenever one of the taught patterns recurs, *both* discriminators will accord it a score of 16 to 16 regardless of its orientation. The cause of this problem is to be found in the fact that the way in which the pixels that make up the grid are connected to the RAMs is similar to the patterns they are required, collectively, to recognize.

RAM 1	RAM 2	RAM 3	RAM 4
RAM 5	RAM 6	RAM 7	RAM 8
RAM 9	RAM 10	RAM 11	RAM 12
RAM 13	RAM 14	RAM 15	RAM 16

RAM arrangement and image patterns coincide

In order to avoid this pitfall, it is standard practice for the designer to connect the pixels to the RAMs in a random fashion. A typical arrangement is shown opposite. The diagram should be read, like a printed page, from left to right and top to bottom. Thus, for example, a two pixel bar down the left hand side of the grid will set address 1000 in RAM one and 1100 in RAM two and a horizontal bar across the bottom will set address 0001 in both RAM one and RAM two. With perseverance, it is possible to construct a table (see page 230) which shows the addresses which have been set in two discriminators arranged in this fashion, one of which has been trained on the four vertical, two pixel bars and the other on the horizontal ones.

1	2	3	4	5	6	7	8
9	10	11	8	12	10	6	4
11	13	6	10	14	9	15	13
2	8	1	5	2	12	16	7
3	9	6	10	7	13	12	1
11	7	15	5	9	3	15	4
16	14	2	8	1	16	14	16
13	3	15	12	5	11	4	14

RAMs connected to matrix at random

If we now look at the response of these two discriminators to a vertical bar down the left side of the grid, it can be seen that in the vertical discriminator it scores 16 out of 16 and in the horizontal one, 11 out of 16. This process can be repeated for each of the eight bars to produce another table (see page 231) which shows the response of the two discriminators to each of the eight taught patterns.

It is now evident that the technique of connecting up the address terminals of the RAMs at random has enabled the system to distinguish between patterns with which it was hitherto unable to cope. This demonstrates both a particular point and a general principle. First, with regard to vision systems, if our single layer net is an accurate, albeit highly simplified, replica of the information processing mechanisms used by the brain, then it has become evident that the precise way in which the rods and cones in the eye (the equivalents of the pixels on the matrix) are connected to the synapses on the neurons that process the information the eye provides is totally unimportant; it does not matter which rod is connected to which synapse so long as the connections as a whole do not form *any* kind of pattern. Randomness is not a symptom of a disorderly system, it is a prerequisite of an effective one. It is also worth pointing out that

RAM number	Horizontal discriminator				Vertical discriminator			
	bar 1	bar 2	bar 3	bar 4	bar 1	bar 2	bar 3	bar 4
1	1000	0100	0001	0010	1000	0100	0010	0001
2	1100	0001	0010	0000	1000	0110	0000	0001
3	0101	1000	0010	0000	1000	0000	0110	0001
4	0000	1000	0000	0111	1100	0000	0010	0001
5	0000	0110	1001	0000	1000	0100	0010	0001
6	0000	0011	1000	0100	1100	0010	0001	0000
7	0001	0000	0010	1100	1000	0100	0011	0000
8	0010	0101	0000	1000	1100	0010	0000	0001
9	1010	0000	0101	0000	1000	0100	0011	0000
10	1000	0011	0100	0000	1100	0010	0001	0000
11	0110	1000	0001	0000	1000	0100	0010	0001
12	0000	0001	1100	0010	1000	0110	0010	0001
13	1001	0000	0010	0100	0000	1100	0010	0001
14	0100	0000	1000	0011	0000	1000	0000	0111
15	0000	0011	1000	1100	0000	1000	0110	0001
16	0100	0000	0010	1001	0000	1000	0000	0111

		Vertical discriminator	Horizontal discrimination
Vertical	Bar 1	16	5
	Bar 2	16	10
	Bar 3	16	7
	Bar 4	16	6
Horizontal	Bar 1	7	16
	Bar 2	4	16
	Bar 3	5	16
	Bar 4	7	16

it follows that, even if one witnessed two nets operating with equal efficiency, it would not be possible to infer that their construction was precisely the same.

The general principle which is illustrated by this particular case is that it cannot be assumed that randomness, or lack of organization, in natural structures is some kind of 'second best' solution that would be improved by the kind of orderly architecture which we are, by training, predisposed to consider superior. On the contrary, the lesson would seem to be that randomness is something that nature finds easy to handle and puts to very careful use. The engineer who aims to emulate natural systems will be well advised to study the uses of randomness and turn them to his advantage.

Incidentally, the rather laborious calculations that are necessary to compile tables like those above, even for a very simple single layer net, illustrate the cumbersome processes that a computer has to go through in order to simulate the behaviour of a net. If one imagines the size of the tables that would be required to catalogue the responses of a net covering an image containing thousands of pixels, it can immediately be seen why the task of compiling them exceeds a computer's capacity to complete it in real time.

Happily, the availability, off the shelf, of RAMs with a capacity of up to 64,000 bits makes it possible to investigate the behaviour of nets of a size and complexity which would defy any process of computer simulation. Since, as has been explained, the RAMs are guaranteed to behave in the same way as the nerve cells of the brain, a net, even a single layer net, containing an appreciable number of them can plausibly be called a facsimile (though obviously a simplified one) of some small part of the brain and might seriously be expected to achieve some of the things which the brain can do.

It must be emphasized that using a net does not involve the use of any top-down model of intelligence; the machine does not have to be provided with any program or algorithm; it is simply connected up to a source of information, in our case a visual image.

As a result of the efforts of one of the authors and his colleagues at Brunel University, a sizeable single layer net has been in existence since 1981, and the results that have been achieved go a considerable way towards confirming the belief that machines of this kind can do things that a conventional computer finds difficult to match. This is not to say that a neural net is 'better' than a computer or a substitute for it, but it may well be taken as evidence that bottom-up methods have as much of a contribution to make towards the construction of 'electronic brains' as top-down ones.

The machine in question has been christened WISARD (for Wilkie, Stonham and Aleksander's Recognition Device) and a brief history of its development is given in Appendix I at the end of this book. This account is of interest, if only because it shows very clearly how indispensable to the whole project was the revolution wrought by the silicon chip. WISARD is vastly larger than the systems we have discussed so far: it operates on images with a matrix of 512 × 512, or 262,144, picture points, and each of its discriminators is made up of 32,768 RAMs each with eight address terminals, and, therefore, a total capacity of eight million bits.

One of WISARD's direct ancestors, a much smaller machine called MINERVA, which was completed in 1972, had already achieved remarkable success in recognizing handwritten characters, far surpassing the performance of computers with infinitely larger capacities, which were limited to the familiar 'machine readable' codes used on credit cards and cheques. When WISARD was finished in 1981 it was subjected to what must surely be one of the ultimate tests for any pattern recognition system – the recognition of human faces. It achieved one hundred per cent success, discriminating consistently between 15 different individuals (and it is important to note that the machine is taught and tested on live images, not photographs or other static pictures) and even exceeding the expectations of its designers in so far as it is sensitive to the expression on a face as well as its identity, and can be trained to distinguish different expressions on the same face as well as different individuals. Nor is it disconcerted by changes in lighting, the complications introduced by facial contortions such as exaggerated grimaces, or even simple 'disguises' like spectacles or false moustaches. This is, of course, a consequence of the system's ability to generalize.

In order to maximize its value as an experimental machine and to fit it for the widest possible range of useful functions, WISARD is so constructed that its total storage capacity can be utilized in different ways. For instance, instead of reducing the pixels in the whole 512×512 matrix to starkly contrasted black and white, it can be reorganized, at the flick of a switch, to concentrate on a smaller area and distinguish the more subtle variations of tone. Or, if an intelligent framestore (see page 182) is used to select the significant features of an image, WISARD will concentrate on these and ignore the remainder of the picture.

WISARD is now commercially available as a pattern recognition device for robot and other applications. But it will obviously be necessary to press on well beyond the level of complexity represented by a single layer net, however large, before we can expect to capture any genuinely intelligent qualities in this kind of machine. In order to discern in which direction the advance should go, and what its next objectives should be, we might usefully examine some of the drawbacks and limitations of a single layer net of silicon neurons such as WISARD.

The most striking fact about a single layer net of this kind is that, despite its impressive ability to recognize complex and subtly different patterns, it has no 'mind' in the sense that we defined the word in Chapter Seven. For all its 'cleverness', WISARD does not have a state structure, indeed it cannot really be described as an automaton at all. The only level at which its behaviour can be described in terms of automata theory is that of the individual RAMs, and each of these is merely a trivial automaton, having one input (a taught pattern) and one output (a firing). The RAMs do not change their states in response to an input but only as a consequence of being taught. And the teaching process that WISARD undergoes is quite different from that which was applied to the learning automata in Chapter Seven. In the former case, teaching is a means of changing the state of some of the trivial automata that, together, make up the machine. In the latter, teaching enabled the automaton to learn not only new states but also the rules that governed the transition from one state to another, a transition which it was subsequently able to perform for itself in response to an input.

One of the objectives we might reasonably set ourselves, therefore, is to endow a WISARD-type machine with a state structure – preferably a learned one rather than a ready made, preprogrammed one, for the latter would involve constructing a model of a particular kind of intelligence, something that, if we are to stick with the bottom-up approach, we would clearly wish to eschew.

If the effort to transform a single layer net into an automaton with a capacity to learn new state structures was successful, it could reasonably be expected that the combination of a learning automaton, having the ability to 'think' inductively by reference to its experience, with the net's inherent talent for discriminating between and generalizing about extremely complex experiences in the form of elaborate patterns would prove very powerful. The resulting machine would have, at least in principle, the potential to outflank two of the three main obstacles which have bogged down the preprogrammed, top-down approach to artificial intelligence.

Turning from ends to means, the most glaring difference between the organization of a single layer net of silicon neurons and the natural nets in the brain is that the former lacks any provision for feedback, something which, as we have seen, the natural system displays in bewildering profusion and employs to keep all the neurons informed about the behaviour of their fellows. A natural net is thus a system which acts in concert rather than an assembly of separate devices each reacting to inputs in isolation. One might therefore anticipate that, if a single layer net is indeed a first step towards a real artificial brain, the next stage would involve the provision of feedback loops of one kind or another which would transform it into an interlinked net.

Lastly, it may well be felt that the 'teach' mechanism which is necessary with a machine of the WISARD kind is essentially 'unnatural' and, thus, undesirable in a device constructed on bottom-up principles. If it could be dispensed with we might feel, almost instinctively, that the whole artificial system had moved closer to the natural one.

11. The Intuitive Brain

And Thought leapt out to wed with Thought
Ere Thought could wed itself with Speech

Alfred, Lord Tennyson, 'In Memoriam'

In discussing how a single layer neural net can be transformed into an interconnected net by the provision of feedback, we shall in fact be following, stage by stage, a program which has already begun. The main aim of this project for refining and developing WISARD's abilities is to discover and test the 'emergent properties' of machines of increasing complexity and sophistication. Since the fact that they possess emergent properties is the most important and significant factor differentiating neural net structures from von Neumann computers, it may be as well to start by defining what we mean by the phrase.

The three most distinctive properties of WISARD are its ability to recognize images and assign a variable degree of confidence to its identifications; its ability to generalize; and its sensitivity to small differences between different images coupled with an ability to recognize these at a single pass (unlike a computer system, which, if it could distinguish such variations would have to compare the patterns it perceived with a whole range of stored templates in order to classify them). All these properties are, as should be clear from the previous chapter, a consequence of the machine's structure – any machine built to the same design will have the same properties.

A computer in itself has no emergent properties at all. The properties it enjoys once it is programmed are very considerable, but they emerge from, or are a consequence of, the program not the machine. The assumption of the top-down school is, of course, that it is possible to provide more valuable and useful properties by constructing a set of rules, or a program for a universal machine, than by building a specific kind of machine. Against this, a supporter of the bottom-up approach can advance two arguments. First, as has been demonstrated, a machine like WISARD has emergent properties which it would be quite impractical to encompass in the form of programs, but which seem to resemble some of the properties of the brain. The second point follows from this. If we wish to understand the emergent properties of our own brains, and we find it impossible to do so by designing programs which have the same properties, the only other course open to us is to construct machines that display those properties. The force of this argument can only be avoided if it is assumed that the brain, like the computer, has no emergent properties which are attributable to its structure and which result from its experience; the evidence that this is not so seems more than ample.

If the unconnected RAM-neurons in WISARD are to be linked together by feedback loops, the obvious, perhaps the only, way of doing so would appear to be by wiring up the outputs of some RAMs to the inputs of others. The consequence of such a course of action would be a machine which behaved in ways that were not only complicated but also exceedingly difficult to follow. Fortunately, there is an ingenious intermediate solution which allows us to keep track of what is happening. The output of each discriminator, it will be recollected, consists of a simple number, the total number of RAMs which are firing in response to the image being displayed. It might, for example, be that 20,000 RAMs out of the total of 32,000-odd are firing in one discriminator, while in another only 10,000 are firing.

The responses can be very easily translated into the form of a bar graph, such as is used to represent statistics of all kinds. This graph can be then displayed in the form of a television image that changes with the responses in the discriminators. In the case we have taken the screen would look like this:

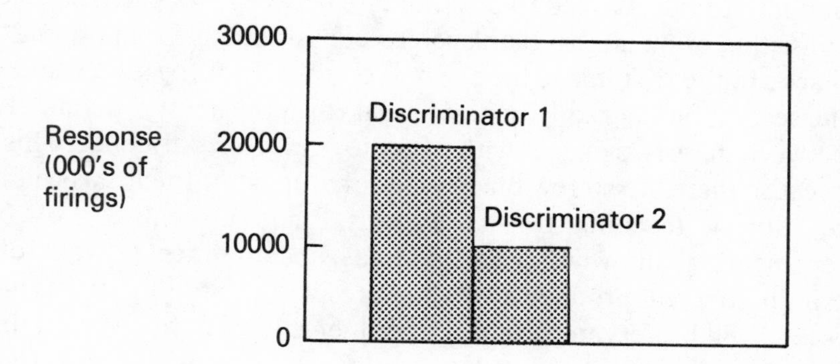

Responses displayed to provide feedback

The result is that all the responses of all the discriminators to one image are represented in graphic form, as a pattern. The feedback is achieved by using a second camera which translates this pattern into digital information. The pattern of the responses is mixed with that produced by the image itself and fed back into the net. In fact, there is a delay built into the arrangement so that the feedback image that results from one scan of the graph is mixed with that produced by the next scan of the image. The arrangement is shown in diagrammatic form below.

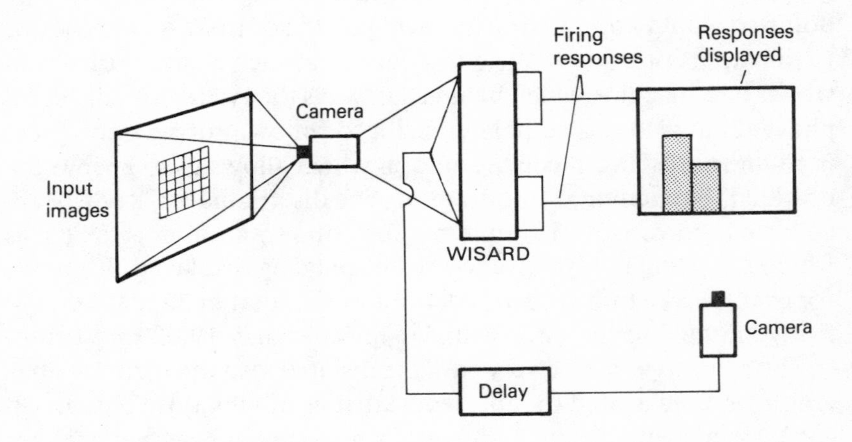

WISARD with feedback loop in place

What takes place when this new system is brought into action? Suppose that it is asked to discriminate between two patterns, each consisting of an almost blank screen containing one small speck of black. In the first image the speck is on the left of the screen and in the second on the right. In its original form the net will distinguish between these two images with a very low level of confidence. This is because there are so few black pixels that most of the RAMs will be set during teaching to the address made up entirely of 0s, representing all white pixels and, as a consequence, both discriminators will produce a very high response to both patterns and results will be very uncertain. Experiments have in fact shown that WISARD distinguishes such patterns with only 30 per cent confidence. One would obviously wish the machine to make a distinction which we ourselves achieve very easily with a much higher degree of certainty.

If the new machine, with the feedback arrangements we have described, is taught these two images the result is quite startling. To understand what happens let us follow the process of teaching one of the discriminators to recognize one of the images (the other discriminators will of course be switched off, or not 'enabled', in order to prevent them learning the same image). The first time the image with the speck is scanned virtually all the RAMs will be set to store a 1 at all 0 addresses except those few which 'see' the black pixels making up the speck. On the next scan, the set RAMs will continue to be addressed and will, therefore, output the 1s they contain and this output will show up as an image on the screen which monitors the response. Since the system recognizes the image with almost total confidence, the response graph will soar up the screen. The third pattern of information received by the net will now consist of a mixture of the pattern made by the speck and that fed back from the output – in effect the discriminator is now 'seeing' both the original pattern *and* the pattern produced by its own response to it. The result is a larger and more varied combined pattern which resets many of the RAMs in a much more distinctive pattern.

When the teaching process for both discriminators is complete and the system is asked to recognize one of the two specks and say how confidently it does so the effect of introducing feedback becomes evident. Again, after the first couple of scans, each discriminator is seeing the image and, this time, the pattern produced by the responses but, as a result of the teaching process, each discriminator is set to recognize only its own response and not that of its fellow. The result is that once a cycle of input, output and feedback is completed the discriminator which recognizes the image 'votes' for it with a higher number of firings while the discriminator which has learned the other image casts fewer votes.

The mechanism at work is not difficult to understand. Take the discriminator that has been trained to recognize the image that is being shown. At first it sees only the speck, and responds with a few firings (the speck alone resembles the mixed pattern which has been learned only marginally), but next time round it sees both the speck and its own rather tentative response to it. As a result many more of its RAMs fire, which in turn creates a response which improves the match with the learned mixture of the two patterns and so produces a better response, which in turn creates a better feedback image, and so on. Exactly the reverse process takes place in the other discriminator, causing it to become less and less confident that it has ever seen *this* speck before. The build up of confidence is very rapid; within three or four cycles WISARD will now distinguish the speck pattern with 90 per cent confidence.

The whole process is analogous to the mechanism which fixes the starting prices for a horse race. If past form causes a preponderance of gamblers to bet on one horse then, in order to balance their books, the bookmakers will respond by shortening the odds on that horse, causing it to become a favourite and, therefore, to attract more punters, while the odds on other animals will lengthen as a consequence of the fact that few wish to bet on them. In fact, of course, this is a result of a closely analogous feedback system, whereas the totalizator (where the gambler is quoted no odds but, if he wins, simply accepts his share of the total cash taken on the race in question) is a system in which the feedback element is eliminated.

The behaviour of the machine that has been equipped with feedback of this kind is very reminiscent of that of a human being who, faced with an image which he thinks he recognizes, takes a second look to make sure. Take the case of someone who, flipping through a book, sees a picture which he feels is familiar. His attention, which had previously been devoted to absorbing new images, is now concentrated on the task of making sure whether the picture he is looking at is really the one he knows or simply very similar to it. In effect, having made a hypothesis he now tries to confirm or reject it. In the case of a human being, of course, what is fed back is the inner image of the remembered picture, not simply a crude measure of the response that is being obtained, and we shall be seeing how this mechanism can be mirrored more closely in a moment. But, in the meantime, it should be clear that, even at this point, we have produced a mechanism for establishing hypotheses and then checking them which is very similar to, if not identical with, our own behaviour.

The way in which feedback is provided, by the use of a visual display unit and TV camera, though extremely valuable for experimental purposes, hardly looks like a model of any natural system and it is important to recognize that it is only a means of obtaining a logical structure which could be achieved by additional wiring circuits. What is, in effect, happening is that the individual RAMs in each discriminator are being given the means to influence each other's behaviour. The same effect would be achieved if the output terminals of all the RAMs in a discriminator were linked up to the address terminals of all their fellow RAMs, including their own. For what the display screen and camera are doing is creating an image representing the output of the whole discriminator and feeding it back to the address terminals.

The interesting thing is that, apart from the improved performance that is achieved, this arrangement, however it is realized in practice, begins to endow the discriminator with a state structure. Rather than being merely a collection of separate and trivial automata, the net has become an automaton in its own right. The fact that the pattern that is being fed back is delayed by one step gives the system that internal concealed state which is the criterion of an automaton. The feedback is a consequence of one input and determines the next output, or, to put it another way, it represents a sequence of 'if . . . then . . . ' instructions to the machine. We can, therefore, draw a state diagram for it. If, for the sake of simplicity, we say that, when a pattern is recognized, the response of the discriminator goes up in steps from 100 to 500 to 600, etc, then the state diagram overleaf will describe the automaton. Since the output and the state of the machine are identical (both consist of the number of firings that are occurring), and the input is simply the taught image containing the speck (the feedback is *not* an input to the system, as becomes clear if we think of the arrangement in which it is provided in the form of 'hard' wiring), the 'if . . . then . . . ' statements take the form: 'If (state) and (input) then (next state).'

The rate of firing does not, of course, continue to increase indefinitely. After the sudden burst caused by the arrival of feedback it will quickly stabilize and the automaton will settle into a state which will remain constant so long as the original image continues to be input. But one further factor emerges from this first level of improvement. It might be thought that, when the input is removed, the automaton would immediately revert to the neutral or random state. But this is not so, for it will be remembered that the feedback is delayed by one step (this would be so in any electronic system, natural or man made, for what causes the transition from state to state is the feedback and, since each state must have some duration, it follows that State 600, which is brought about by the feedback created in State 500, must actually succeed it in time). The consequence is that even when the original speck image is removed the net continues to respond to its own response and this will persist,

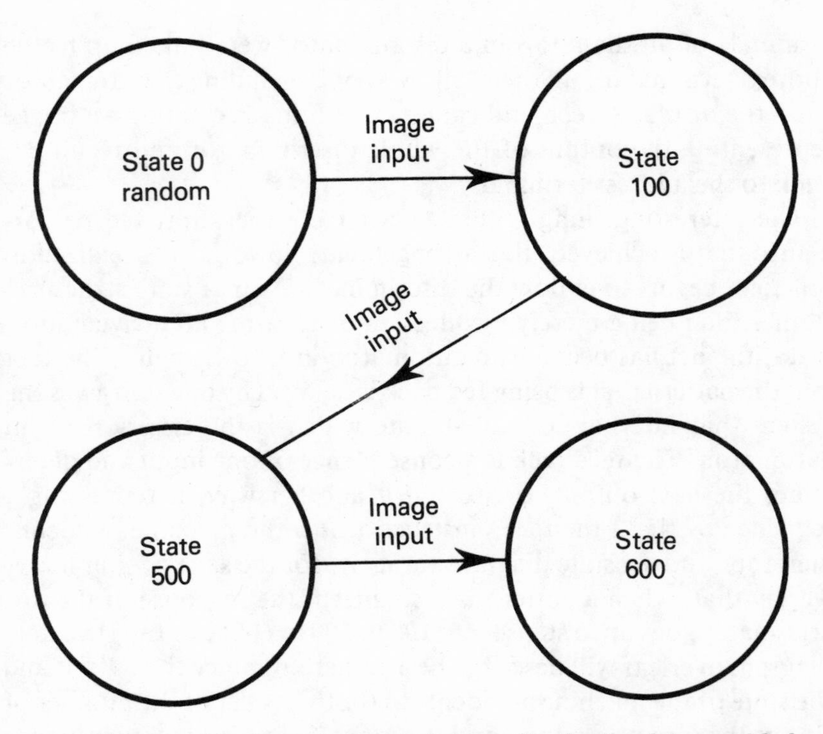

State diagram showing response to image after feedback is provided

with a decreasing degree of confidence, until the 'memory' eventually decays altogether. The continuation of the state structure above, assuming that the speck image is removed at State 600, will therefore consist of a succession of further states with fewer and fewer firings as shown opposite. The level of response will naturally drop most precipitately to begin with, when the speck's contribution to the input is lost.

The machine as a whole now has the property of remembering, in the short term, *which* image it last saw. Nice though this may be, it is not in itself a very significant advance (a computer, after all, could remember, and list in order, hundreds of thousands of inputs). What would be much more startling would be the discovery of some way of enabling the machine to memorize not just which image had been seen, but the image itself. This is the objective of the next level of sophistication.

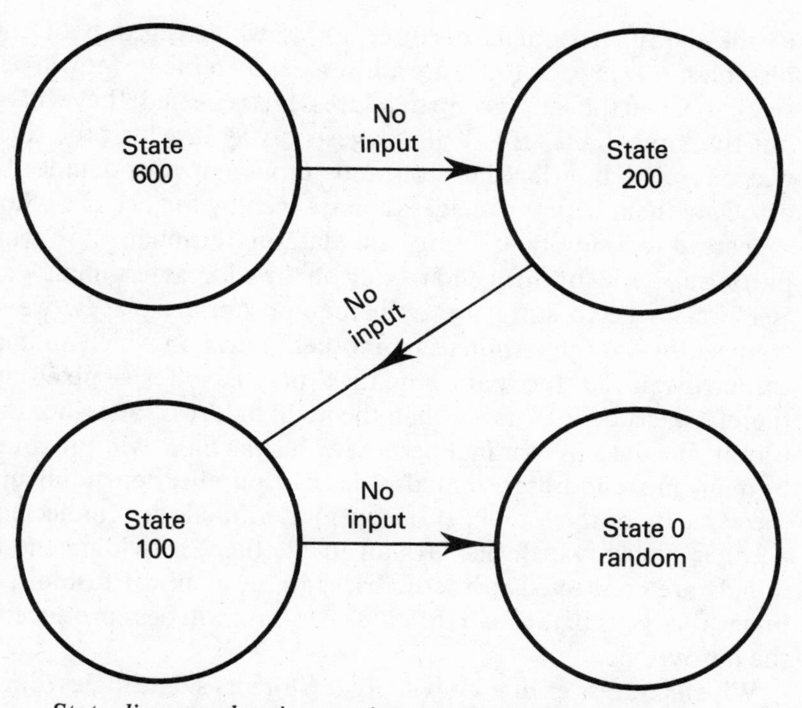

State diagram showing persistence of short-term memory

The first step towards creating an ability to remember an actual image, as such, is to find some way in which the net could be persuaded or adapted to store and output such a thing – the difficulty being that, in the arrangements we have examined so far, the input goes to the address terminals which do not tell the net *what* information to store. They simply determine *where* information should be stored. Moreover, if one wishes to retrieve or recover an image that is being stored, the way that the output has been organized is singularly unhelpful, for it only reveals how *many* RAMs are firing, not which ones. To see how these problems can be overcome, we must go back a couple of steps: to a net which has no feedback and which has only one discriminator.

The terminals which determine what a RAM stores in a computer are, as explained in Chapter Ten, the data-in ones, which we have adapted to serve the purpose of teach terminals, telling the RAM when to learn rather than what to learn. It will, however, also be remembered that inputs at the data-in terminals are directly correlated with the output – if the digits 101, for example, are present at the data-in terminals of three RAMs when a group of three bistables, one in each RAM, are addressed, then the RAMs will be so set that, when the same group of addresses is repeated, they will output the same digits, 101. This suggests a way in which the silicon neurons might be utilized to store and output images and, indeed, to associate them with the images that are seen by the net as a whole.

Instead of simply treating the data-in terminals uniformly, putting a 1 on all of them and relying on the address terminals which 'see' an image to store the 1s in the appropriate places, we can connect the data-in terminals to another matrix on which an image can be displayed (the 'data-in matrix' may have fewer pixels and, therefore, cruder definition than the main 'address' one since there is only one data-in terminal per RAM whereas there will, of course, be many more address terminals, the exact number depending upon the capacity of the RAMs). If the output terminals are connected to another matrix which displays an image then, providing the terminals are connected to the matrices in an identical fashion, the image that goes in through the data-in matrix will be reproduced on the output one.

Which address each RAM will use to store its single digit from the data-in matrix will be determined by the pattern of digits at the address terminals of that RAM which will, in turn, be determined by

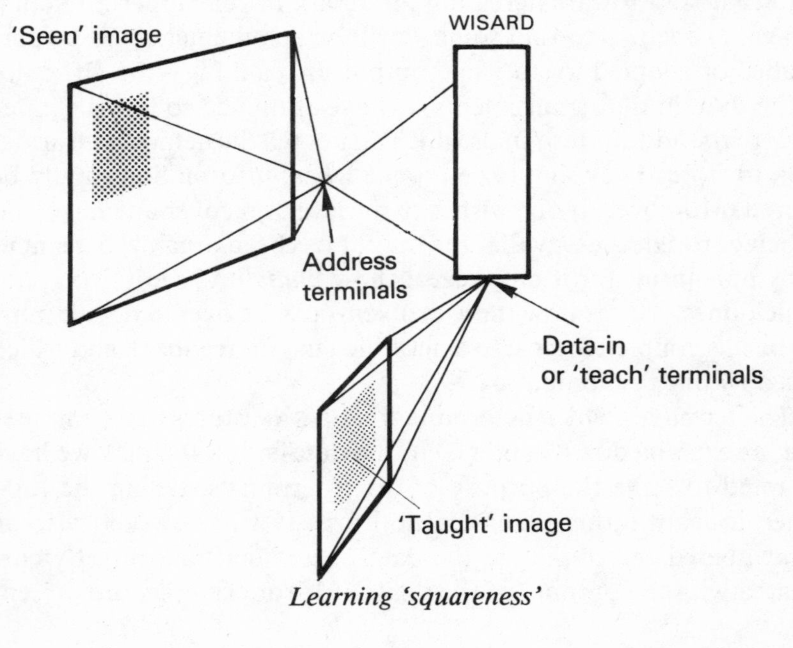

Learning 'squareness'

what image the net is 'seeing' on the address matrix while it is 'learning' the image on the data-in matrix. A system which is taught in this way enjoys some extremely interesting properties. Look now at the diagram opposite. This shows a system being 'taught' a large square which occupies most of the data-in matrix at the same time it is 'seeing' an image containing a small square. The image of the large square is therefore being stored in the group of addresses which is selected by the image of the small square. The process could be repeated with, say, a large triangle on the data-in matrix and a small one at the address matrix. At the completion of its training the net has become an automaton with a learned state structure of a highly significant kind. Its state diagram is shown below, and it can be seen that showing it the image with the small square now puts it into a 'square state' in which it will output the image of the big square, while showing the small triangle will cause it to output the big one.

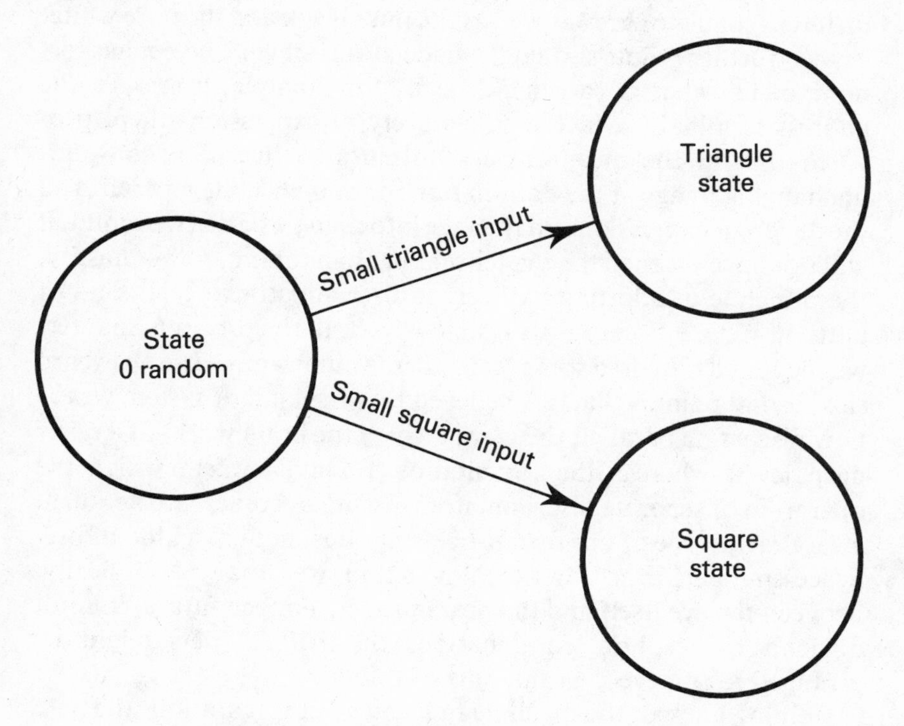

State diagram for WISARD with internal image

The structure could be enlarged as far as one wished; the machine could be taught to associate circles with circles or spirals with spirals, etc. It should, however, be noted that the associations are quite arbitrary. It could equally well be taught to output a square when it saw a triangle and vice versa. The real significance of the state structure is twofold. First, the net now identifies a 'seen' pattern by the internal, or state, pattern it has been taught to associate with it rather than simply by the score, or number of firings, it generates. Second, and following on from this, separate discriminators are no longer needed for each image: one discriminator can remember a whole range of images and recognize them in terms of the output patterns they bring about. In the case of WISARD, this means that we are free to utilize all the storage capacity as one giant discriminator and, therefore, store more information about each image, perhaps allotting several address terminals to cover each pixel, or using the spare capacity to discriminate between different shades of grey as well as the pure black and the pure white.

Two further points should be made at this stage. The net has lost none of its capacity to generalize; if, for example, it sees, at the address terminals, a face with a variety of expressions then, provided that in the meantime it continues to 'learn' a constant, unchanging image at the data-in terminals, it will still generalize all the different expressions on the face into a single pattern. Second, it will not have escaped the reader that in abandoning, rather hastily, the principle of allotting a different discriminator to each kind of pattern, we are discarding a system which, in the previous chapter, we suggested mirrored some reality in the human one. The answer to this second point is that it is believed that the kind of system we are now discussing parallels the way in which the brain works at a fairly deep level, whereas the 'orientation finding' system which the operation of separate discriminators resembles comes into action at a very early stage of our own processing. It is, in fact, a kind of preprocessing performed by neural nets that are situated, physically, between the eye itself and the brain proper. For the future, it is not difficult to see how combinations of artificial nets might be arranged to achieve the same sort of thing.

Coming back to the machine in its latest configuration, the next step is fairly easy to see. We now have an output image which is associated with, or summoned up by, an input image in exactly the same way as, earlier on, the bars on the TV graph were associated with the input images. If we now feed back the output image and mix it in with the input one we shall achieve exactly the same 'confidence-improving' effect. This time the improved confidence takes the form of a clarification or refinement of the internal image or state; each feedback cycle 'fixes' the internal state more firmly. We can also simplify things by eliminating the separate teach or data-in image. This can be provided by connecting the teach terminals directly to the input image. They may 'see' it in a cruder form than the address terminals since there are fewer of them, but this is relatively unimportant – in fact, as we shall explain shortly, the direct correlation between 'learned image' and 'seen image' is important only to the human user; to the machine itself it is irrelevant whether squares are associated with squares and triangles with triangles.

But keeping the two images the same does enable us to see that the machine now has a new property. Suppose that it is shown a face. In its new configuration it will automatically learn the image, for everything it sees is providing information simultaneously to both the teach and the address terminals. It has, in fact, become a compulsive autodidact. Moreover, since the pattern received by the teach terminals is now being fed back, one step later, via the output terminals and mixed in with the input, the image itself is constantly circulating round the system. In fact, when the machine sees the face of John it is 'thinking John'. Furthermore, since the feedback image persists even when the input one is removed, it will, in the short term, remember 'John'. This new state of affairs can be represented diagrammatically as below.

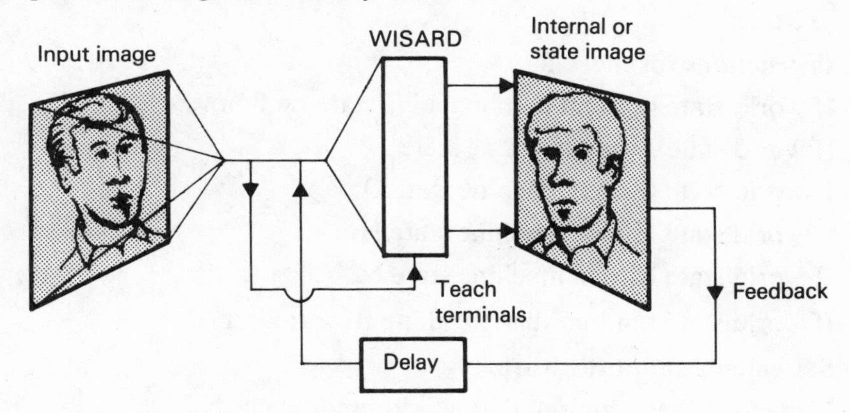

WISARD with internal image feedback

But we have by no means explored all the abilities of this new machine. For providing it with feedback has turned it into a true automaton and, moreover, it is a learning automaton. Even in the early stages, before feedback was added, it has the ability to learn and follow an 'if . . . then . . . ' instruction. For example, it learned the sequence 'If small square input then large square output and state.' It requires, let it be noted, none of the paraphernalia of X = Y mechanisms to be supplied from outside in the shape of a program, for the construction of the 'if . . . then . . . ' statements was an emergent property of the machine itself. However, at that stage of development, once an image was removed the machine reverted to a random state and only moved into a new learned state when it saw another familiar image. Now with feedback added it has new and extremely exciting properties.

Imagine that the machine is shown the letter J. It will proceed to learn a J state, refining it as feedback hardens its level of confidence. But if this is immediately followed by the image of an O, then the machine will perceive at its inputs not only the new image but also, mixed in with it, the feedback memory of the J. Its internal state will therefore change from J to O through a short sequence of intermediate states. If the process is continued with the letters H and N, then the machine will, in effect, have learned a state structure for itself and for the world. If the 'if . . . then . . . ' instructions for the two automata were set out (and remember that this can only be done in theory; inside the machine the instructions do not exist as such but only as a logical consequence of its construction) they would look like this:

Instructions for machine

If world state random then machine state random.

If world state J then machine state J.

If world state O then machine state O.

If world state H then machine state H.

If world state N then machine state N.

If world state random then machine state random.

Statements about the world

If machine state random then world random OR J.

If machine state J then world state O.

If machine state O then world state H.
If machine state H then world state N.
If machine state N then world state random.

Combining these into a state diagram gives a structure like this (note that we do not need to show the transitions as accompanied by inputs or outputs. The world's output, which is also the machine's input, is the same as the world's state, and the machine's output is irrelevant to the world since it has no influence on it.):

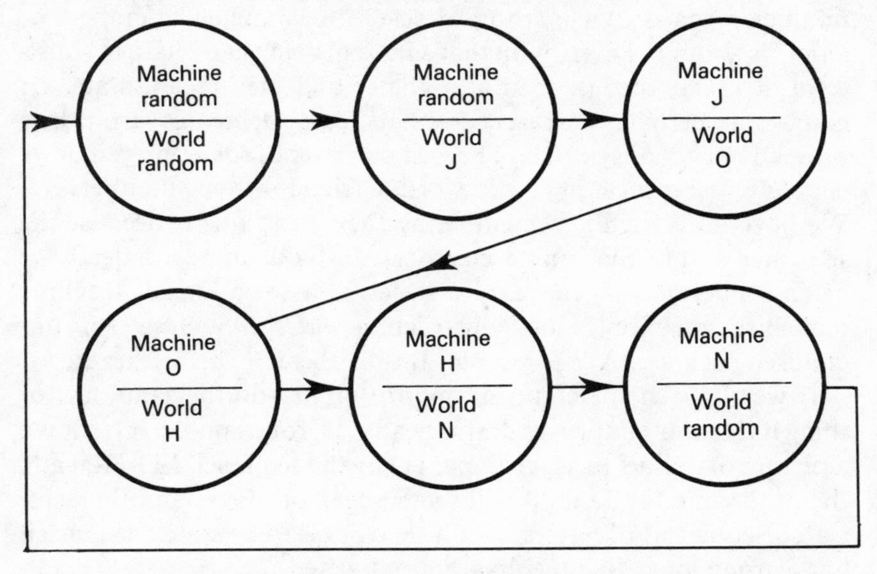

State diagram for learning a sequence of letters

As a result of learning this state structure, the machine now has the property that, when it is shown the letter J, it will automatically traverse the remainder of the structure, outputting J-O-H-N. Moreover, if it is shown an alternative sequence, say JANE, then it will also learn the alternative state structure. In this case, showing it simply the letter J will cause it to enter a J state in which it will remain uncertain, until it sees the next letter, whether the sequence is going to be JOHN or JANE.

These two state structures will, however, be very weakly implanted in the machine, for the fact that seeing, say, a J and then an O leads it to expect (according to the statements about its world automaton) an H begins to drive it into an H state; but unless the H actually appears it may have difficulty continuing on to the N. The world will have failed to do what it was supposed to do, and instead of the instruction for the machine automaton being 'If machine state O and world state H then machine state H', it is confronted with one that goes 'If machine state O and world state random then machine state (?).'

This inability to remember what is, to the machine, an arbitrary sequence of images mirrors a very similar human weakness. One of the reasons for the 'magical' properties often associated with the number seven is that a string of seven 'meaningless' numbers or letters is about the maximum that a human being can retain in short-term memory, that is, read at a glance and then recapitulate. Of course, as people, we rarely need to learn strings of arbitrarily ordered images or symbols. The vast majority of such things that we encounter have meaning – as a word, a friend's phone number, etc. We have no difficulty remembering these, and this is because the sequence and its meaning are associated in our minds; indeed, we often deliberately try to create such sets of associations by teaching ourselves mnemonics or constructing *aides mémoires* for our children such as, 'A is for apple, B is for bear, C is for cat, etc.'

If we allow the machine an opportunity to do the same sort of thing its memory improves dramatically. If, for example, it is shown a picture of John's face accompanied by the sequence J-O-H-N and then the same for Jane, it will remember both faces and the associated sequences of letters with little trouble. The state structure it has learned looks like the diagram opposite.

Once the machine has learned this structure it has the property that, when shown a picture of John or the sequence J-O, it will go through the sequence associated with it: outputting J-plus-John, O-plus-John, etc and will do the same for a picture of Jane or the sequence J-A.

The reason for WISARD's success with these images or sequences which are associated with one another is relatively easy to understand. The property of generalization means that, once switched into a 'John and J' state by the appearance of one or other image, the persistence of the face or the continuation of the sequence gives the machine an input from the world (or a state of the world) which is close enough, in terms of the internal states it brings about, to what is expected to fall within the generalized patterns which represent

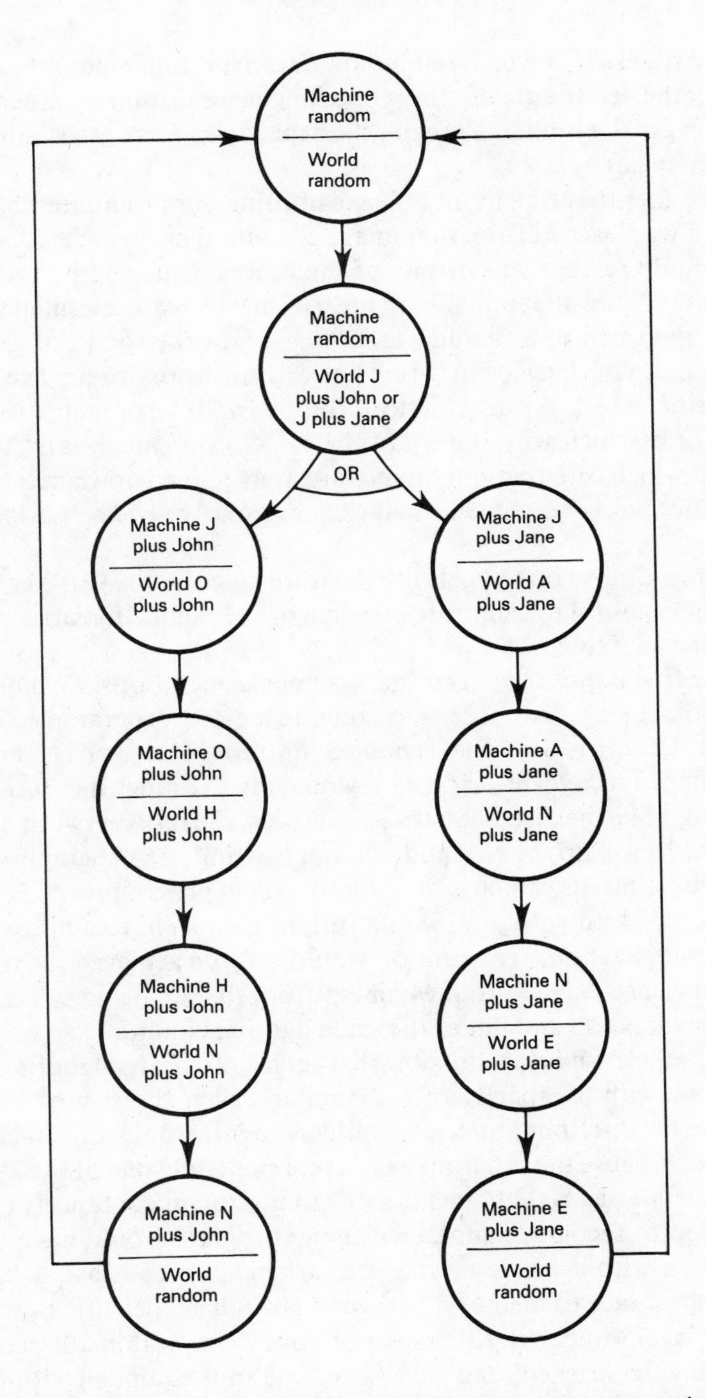

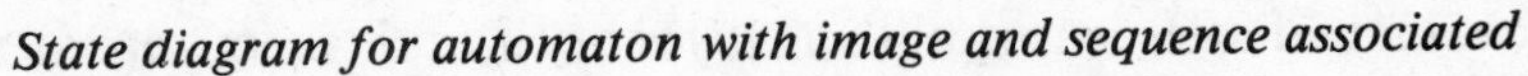
State diagram for automaton with image and sequence associated

those states. It is worth reminding ourselves at this point that, the larger the net, the greater the scope for generalization – or, perhaps, the bigger the brain the greater the capacity to form associations or attach meanings.

The fact that one input image can bring about an internal state which was learned from that image *plus* another, associated, one is uncannily reminiscent of one of the quirks displayed by our own brains. We are all familiar with the way in which quite a small sensation, the smell of a familiar tobacco, say, or the sound of a particular piece of music, can 'bring back' the memory of a place or an occasion which we had almost forgotten. Interestingly, we feel memories which are triggered off in this way very much as a 'state of mind' which often cannot be very precisely pinned down in terms of specifics but is none the less an extraordinarily vivid sensation.

Before we move on to deal with the third stage in WISARD's growth, it may be useful to elaborate on some of the points that arise out of the first two stages.

The first is the economy of the machine's operations. To illustrate the point, let us suppose that some ingenious programmer developed an algorithm that enabled a traditional von Neumann machine to recognize a face. This would involve much measurement of things like the distance between the eyes, the ratio between height and width of brow, etc; and actually writing the program would absorb hundreds of hours of work by skilled programmers. Even if it succeeded, the program would still, of course, have no provision for generalization – recognition would only be achieved if the face was always presented to the camera from the same angle, with the same expression and under the same lighting conditions.

Nonetheless it is just possible that such a program might fit into a memory with a capacity of eight million bits, the size of one of WISARD's discriminators. It would recognize only one face, of course, because the program would represent only one of the $2^{8,000,000}$ possible messages that a memory of that size can contain. If it was required to recognize another face the program would have to be rewritten with a whole new set of variables. In contrast, WISARD commits a face to memory, and generalizes it sufficiently to recognize it again under a wide range of conditions within 20 seconds. And having 'learned' one face it can, once it is equipped with feedback, immediately learn another. In theory it could memorize all the

possible patterns that can be formed by eight million bits – tl
in practice generalization reduces this number dramaticall
saturation would set in at an early stage.

We mentioned earlier that the fact that the teach terminals were connected to the input matrix in a manner which produced an internal image which actually corresponded with the image that was being seen was a convenience rather than a necessity. The need for the internal or 'learned' image to correspond with the 'seen' image arises because the internal image is also the output one – and displaying it on a screen is, therefore, the only way in which an experimenter can know what state the machine is in, or what it is thinking. From the point of view of the machine itself, the nature of the internal pattern which an input creates is quite irrelevant; if seeing an image of a square creates an internal image of a triangle it does not matter in the least – to the machine it will still 'mean' square and will only be recreated by another square.

This may seem to require further explanation, in order to relate it to our own sensations, especially those we experience when we remember an image rather than looking at one that is before our eyes. If, for example, we think of the exterior of a house that we knew well in the past, we have a very distinct feeling of a sort of epidiascope inside the mind, projecting the image of that house on to an internal screen. This makes it very difficult to abandon the idea that the image persists, as such, somewhere in our head. The truth, if the evidence of work with neural nets is to be believed, is rather more subtle.

Our internal image of the house is a highly generalized one and takes the form of a generalized 'state' of some part of the brain created by the images of the house that we saw in the past. When we remember what the house 'looked like', what is happening is that the same state is being recreated. As a result we 'feel' exactly as we felt (or nearly so, depending upon the vividness of the memory) when we actually looked at the house. The pattern of information that our eyes generated at the neural synapses when we saw the house existed only momentarily, but the generalized pattern of feedback information that was created by a succession of images of the house, the internal pattern, can be conjured up, either because some associated pattern (the name of the house, perhaps) triggers the brain off on that sector of its state structure, or because somewhere in the long-term memory a bit of our own neural net has been per-

one neuron calls a series of other neurons

manently 'set' in order to store the image. WISARD, of course, would only have a long-term memory if the teach terminals were disconnected and the functions of its RAMs frozen. But it is easy to see how, if its output was connected to the teach terminals of another neural net, that second net could be set permanently to store an internal image whenever required.

Turning now to a brain, if we no longer care about the precise form which the internal image takes as a matter of principle (though retaining it as an indispensable experimental convenience), it becomes clear that, just as the connections linking the eye's retina to the synapses can, indeed must, be partly arbitrary, the same thing applies to the nerves which lead to the dominant synapses. Since the means we have used to provide feedback, again largely to meet our needs as experimenters, has the effect of connecting every neural axon in the brain to every synapse, we can now say with complete confidence that this collection of neurons is not dependent upon any precise structure of connections – the more random and multifarious the connections of the neurons to the input and to each other, the better.

It can also be seen that we have achieved the objective of constructing a machine which will learn from experience and by induction. Its ability to associate related experiences has provided us with a very clear instance, in the case of the John and Jane images and names, of its capacity not only to learn to recognize things but also to learn what they mean. If the John and Jane examples looked a little like programming of some kind, it should be borne in mind that this machine will, in principle, learn to recognize everything it sees and to associate it with anything else that happens to be in its field of view at the same time. To take just one of the problems that AI found so intractable, if a WISARD-type machine were shown the sequence B-L-U-E twice, once in association with a slice of rare steak and once with a clear summer sky, it would be in no danger of confusing the two. It would have a 'blue meat' state and a 'blue sky' state; just like a human being, its 'blue state' would be one of uncertainty until it was shown in which context the word was being used.

Finally, we return to the subject of the output. As human beings, unless we are graphic artists, we do not produce an output which resembles our visual inputs – when we see an image we do not output the same image. We transform the input image into an internal state image just as the machine does and, if we see it again, or are reminded of it, we may recapitulate it internally, again just as the machine does. But if we produce a response or output when we see an image it is likely to take quite a different form. We may comment on it, run away from it, fall in love with it, or whatever. Clearly, one of the more intriguing prospects for WISARD's long-term future is to connect its output terminals up to another neural net which is able to translate the patterns they carry into these kinds of responses.

This will also be essential if the machine is to gain other kinds of input and broaden its experience. In its present state, as a visual system pure and simple, it cannot know that the world exists in three dimensions, that boxes are hollow, feathers light or steel hard. In order to learn these things it will have to acquire a manipulator with a sense of touch, and so forth. Becoming a robot in this sense of a machine that operates in the physical world may be the next step. Language, which to us seems but a short step away from vision, is a much longer-term prospect. For, if this machine is to master language it will, just like a human infant, have to face up to a painstaking, and possibly extended, learning process, and that process will be of limited use if the machine is restricted to the kind of experience that it can gather as a passive observer of the world. If we want to discover whether WISARD, or some future much enlarged and refined descendant, can indeed think and act like a person, then we will have to start by trying to educate it like a person. The first prerequisite for that endeavour may well be attaching it to a body. Human beings, after all, learn to control their own bodies before they learn to read.

The final stage in the program for WISARD's development is indeed concerned with this problem of enabling the machine to gather better information about the world by giving it the means to control its own movements. But to equip it with an arm would be to leap ahead of ourselves. The first movement it needs to control is that of its own eye. It should be added, in passing, that, so far as WISARD itself is concerned this development is, at the time of writing, still a plan for the future rather than an accomplished fact, though the principles involved have been tried out on smaller machines and shown to be sound.

This third step is, even more explicitly than the first two, an attempt at reproducing a facet of the human system. One of the secrets of the success of our own vision system lies in the fact that we are able to concentrate our attention upon a particular area, a few words or a few lines of text in a newspaper column, perhaps, while remaining aware of what is happening in a much wider field of view, and having the freedom to switch our attention to another sector of it very rapidly. Thus, even if as far as the newspaper reader is aware his mind is totally devoted to the article he is studying, any sudden or untoward movement detected by his peripheral vision will swiftly attract his attention. This facility for interpreting visual information on two different levels simultaneously also allows us to take in a whole scene while selecting from it those features which seem most interesting or significant.

Any machine which is intended to match or replicate the skills and qualities of human eyesight will obviously have to share this facility to see both the wood and a particular tree at the same time. In the case of the human system, the actual switching of attention from one area to another is achieved by moving the eyes themselves. We are very conscious of this mechanism, even in others, for we often complain, even when we are face to face with another person, that they are 'looking straight through' us. No movement of their head has betrayed their lack of attention, just a movement of the eyes. Thus simply moving a camera about to look at different things would not achieve the same effect; what is needed is some way of 'zeroing in' on any part of an image while still keeping the whole image in view. Moreover, anyone who has been subjected to home movies or video films will know how easily a picture is confused and rendered unintelligible by rapid camera movements, and our brains contain some extremely complicated mechanisms to ensure that, when we move our eyes quickly from place to place, we are not affected by 'camera shake'.

In the case of a machine, it would be extremely difficult to find equivalents for this stabilizing machinery and, to begin with at least, the idea is to achieve essentially the same results as the eye by rather different means; for, in terms of the workings of the visual system, the interesting thing is clearly the control of a closely focused area within a larger image. The use of our old friend the intelligent framestore (see Chapter Nine) allows this coexistence of two images on a single matrix; in effect, the whole matrix is seen at one level of definition while a small 'window' within it is simultaneously seen in much higher definition. The effect is very similar to that which we experience if we hold a magnifying glass over one section of a page – attention is focused on the words that are magnified but we are also aware of the rest of the page. By the use of well-established techniques, very similar to those used on television when a background image, a rock group perhaps, is displayed simultaneously with a part of it, say the face of the lead singer, in close up, it is quite feasible to move the highly defined area around within the boundaries of the larger image.

The technicalities of the thing are not especially important. The crucial question is, how could a machine be taught to move its attention from one part of an image to another when it detected something 'interesting'? Clearly, feedback of some kind is involved in our own case. When we see something 'out of the corner of the eye' and focus on it the action must be the result of that part of the brain which detects a significant object or movement within the overall field of view instructing the eyes which way to move. In order to keep the explanation as simple as possible in the case of WISARD, we shall ignore the feedback systems which we have looked at previously and concentrate on a new one, the purpose of which is to cope with this function known, in the case of human beings, as 'saccadic eye movement'.

The starting point is the pattern at the output terminals which represents the combined pattern of both images, the overall one and the highly defined window. The output can be connected to a control mechanism that moves the high resolution 'window' around the frame – as we have said it does not much matter how this movement is achieved, the important thing is how the system can be taught to operate the controls under its own steam. The teaching is done by means of a joystick exactly like those that are used in many home computer games, the joystick being connected to the teach terminals of the net. (The teach terminals will also 'see' the image itself at the same time.) The whole arrangement is shown in schematic form overleaf.

If the system is presented with an image containing a number of spots, the joystick can be used to move the window from one spot to another in some definite order, clockwise perhaps, around the matrix. The way in which the machine learns is exactly the same as that which was described for a 'learning' robot in Chapter Seven. In effect, we have created a learning automaton which has been taught that the world will present it with inputs, in the form of blobs on the image, and that it must respond with outputs by moving the window to the blobs. When a similar arrangement was set up, using a much smaller net than WISARD, it was found that the system quickly learned to follow a line, or trace the perimeter of a shape, with its window. To start with, if it was trained only on squares, it would quickly give up when faced with other shapes, triangles say, and remove the window to one corner of the image where it proceeded to 'sulk'. But once it had been taught two or three simple shapes, the generalization properties of the net enabled it to trace shapes of any kind.

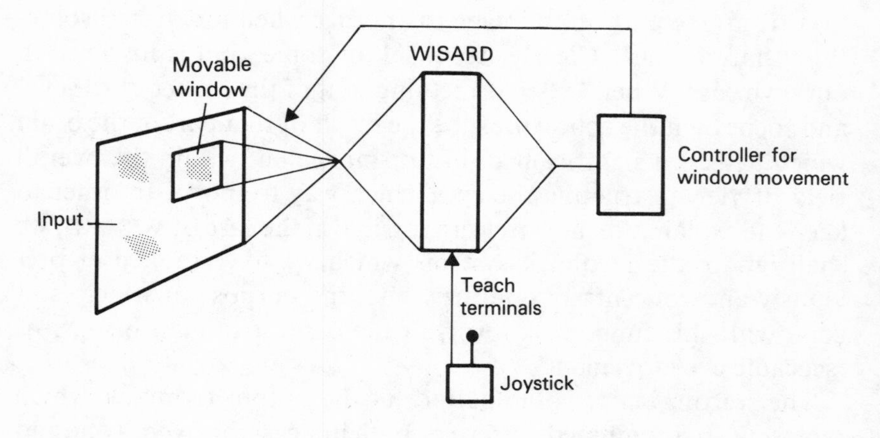

WISARD with window control

To date, we can only speculate about the properties of a system which embodies all the feedback loops we have discussed. It seems likely, however, that it would quickly learn to move its window in jumps from one feature to another and that the kind of labelling mechanism that associated John with J-O-H-N would allow it to learn, for instance, that both the patterns opposite represent triangles; this is the kind of feat of recognition that comes naturally to human beings but, notoriously, eludes algorithmic systems. A combined system should also be capable of developing a capacity for using its window to resolve ambiguities, zooming in on areas of an image which it found puzzling.

We have not described the workings of this third stage in any great detail, if only because we hope that the basic properties which enable the net to operate as a learning system are now fairly clear. It

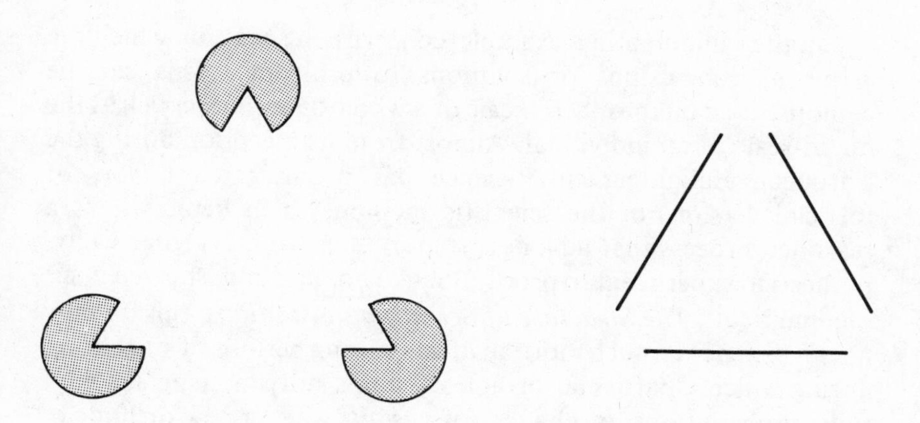

Two 'triangles' that might be understood by concentrating on the significant features

is, however, intriguing that since this experiment was first planned, physiologists have discovered features in the human system that look, in principle, to be very similar. It has been found that feedback loops exist connecting an organ called the lateral geniculate body, the functions of which were hitherto obscure, to the muscles which move the eyeball. It may be that feedback in this circuit of nerves is what 'manages' human eye movement.

One of the strongest reasons, though not the easiest to define, for judging that the neural net is the 'right' model for the brain is the fact that it 'feels' right: it allows us to reconcile our own sense of how our brains work with the way in which a machine works. This contrasts very strongly with our reactions to programmed models, where there is always a sense that, however effective and ingenious the programs, they are not really reflecting a true and complete image of the mind. Take, for example, the programmed methods for scene analysis described in Chapter Nine. If we are not aware of any alternative then we are forced to conclude that our brains must perform some sort of analysis of the relationships between the corners and the sides of a cube in order to recognize it; it does not seem likely and it does not match our own sense of what is going on inside our head, but how else could it be done? The idea that the image of a cube puts some part of our brain into a 'cube state', on the other hand, seems exactly right.

Another illuminating example concerns the way in which we arrive at conclusions or solutions to problems. This can be demonstrated on the macro-scale of a whole discipline as well as the micro-scale of an individual. Almost from its inception during the European Enlightenment, science has promulgated a sort of 'official version' of the scientific method which depicts it as a reasoned process that advances step by step from evidence to hypothesis to experiment to proof. But as Thomas Kuhn showed in his seminal book, *The Structure of Scientific Revolutions*, this is really not so. Science is much more an affair of long periods of stagnation during which a particular orthodoxy holds sway interspersed with sudden revolutions, or changes of *gestalt*, when a new orthodoxy takes over. Nor does the traditional model fit in with the reported experience of scientists themselves.

It is common to nearly all accounts of experimentation or theoretical breakthrough given by the participants themselves that the crucial insight is described as coming not at the end of a long, plodding advance towards the light at the end of a tunnel, but as a sudden and quite unexpected revelation – one moment all is opaque, the next all is light. This seems to fit very well, if loosely in our present state of knowledge, with the kind of sudden 'catastrophic' changes of state which can take place within a neural net. Even more convincing perhaps is that this sensation, which is familiar in the case of everyday problems as well as major scientific breakthroughs, often takes the form of a sudden glimmer which, in a matter of moments, becomes a brilliant glare. This matches exactly the leap from the tentative estimate to the confident decision that takes place when WISARD's RAMs receive the first burst of feedback that confirms its recognition of a dubious pattern.

One final example may perhaps not be out of place. One can see no way in which tying a knot in a handkerchief could help a computer to remember something (unless, of course, its program included an instruction telling it what the knotted handkerchief meant), but one could, even as a practical experiment, show WISARD the image of a knotted handkerchief together with something one wished to remember (it would of course have to be another image) at a later date and be confident that, when shown the handkerchief by itself, the machine would re-enter the handkerchief plus-image state and output the necessary reminder.

If we accept that in constructing machines like WISARD we are beginning to replicate the structure of our own brains, then the first response must be one of humility in the face of the elegance and beauty of the mechanisms involved and the extraordinary intricacy of the original. For it must be emphasized that, even when all the refinements described in this chapter have been added, WISARD will remain an imperfect model of but one small section of the human brain. It will not even reproduce the entire machinery of our own visual system and short-term visual memory, for it has no provision for the preprocessing done by the 'orientation finding' mechanisms described in Chapter Ten or the 'sketch making' one discussed in Chapter Nine. These must be the business of other neural nets with feedback links to the ones which control saccadic eye movements and create internal images.

An ambitious program of research, involving engineers, neurophysiologists and psychologists will be required before we can begin to grasp with any confidence how all the different neural nets which make up a brain relate to each other and cope with the input of the senses and organize the outputs of the body. Yet we may also be permitted to feel some small satisfaction at having reached the starting line, at having found the right model and a means of reproducing it.

Even in the present state of knowledge, the possibilities for enlarging our understanding of our own workings that open up once we accept the neural net model are almost endless, and we can suggest only a few of the avenues that might profitably be explored. One obvious possibility is that of resolving the Chomsky-Piaget debate over the nature of our language using skills. Teaching an entire language to a neural net would be a vast undertaking, equivalent to educating a newborn child, but if it were possible to inculcate the basic elements of language use into a totally unprogrammed machine then the argument would swing very heavily against Chomsky's ideas.

Again, we now have available a very clear indication of what constitutes an instinct. The mechanism that causes newly hatched chicks to open their gapes whenever the light is cut off over their nest, or to freeze when they see the silhouette of a hawk overhead, must be the output of a neural net in which the synaptic weights of the neurons were fixed at or before birth. Similarly, the mechanism that causes a newborn animal to become fixated, in the absence of its natural parent, on an individual of another species must mirror a network somewhere in the brain containing neurons that are permanently set by the input of the first large animate object the creature encounters. It follows that, if one wants to discover where in the brain instincts reside, one looks for areas in which the dominant synapses are absent or inoperative, and if one wants to see which bit of the brain a fixation 'fixes' one looks for a net that sets very soon after birth.

The whole argument over the relative importance of nature and nurture can now be seen as a question of which areas of the brain, responsible for which outputs and inputs, are set at birth and which remain adaptable, with neurons that are susceptible to changes in synaptic weight. A permanently set network is, of course, equivalent to a preprogrammed one and the issue of how far we are ourselves preprogrammed can thus be investigated from two directions. Engineers can construct unprogrammed robots and see how far learning will carry them, and, at the same time, neurophysiologists can identify the areas of the brain that are responsible for particular functions and try to discover whether the neurons concerned have fixed or variable functions.

The prospect of investigating the properties of hierarchies of unprogrammed nets also opens up the possibility of new insights into dreams and the nature of the unconscious mind. Is it perhaps the case that dreams are the result of traversing some areas of a state structure that we are prevented or inhibited from exploring during our waking hours? Is the subconscious in fact the state structure of some vast net or collection of nets the outputs of which influence, but remain unrecognized by, the conscious area of the brain? Will neural nets reach a size and complexity at which they need to sleep in order to sort out and organize the patterns they have absorbed? We know for a fact that the relationships between different areas of the brain are a vital factor in its behaviour – the fact that it is divided into two hemispheres which are linked in a surprisingly parsimonious fashion is but the most dramatic instance of this. It remains to be established which bits control which, how feedback flows between them and how 'superior' functions are enabled to exercise control over more primitive ones. The task of sorting all this out will not be an easy or a short one, but it will be enormously simplified and speeded up by the existence of a kind of machinery that allows us to put hypotheses to the test. The neural net of the future can be to the psychologist and the neurophysiologist what a row of test tubes is to a chemist – a testbed for new conjectures.

Now that we have described a machine built on bottom-up principles and have shown that it is capable, in practice as well as principle, of mastering some facets of intelligence, it is possible to talk in more precise terms about the relationship between bottom-up machines and top-down programs.

It has been demonstrated that bottom-up machines can learn inductively from experience in ways which preprogrammed machines, because of their reliance on algorithms and their lack of the capacity for generalization, cannot. It has also been shown that a bottom-up machine can be transformed into a learning automaton which teaches itself. Leaving aside, until the next chapter, the vital question of why a machine should *want* to learn, it is clear that we can now build a machine, one of the emergent properties of which is the acquisition of experience of the world in which it finds itself.

Taking a bold step, we could then say that, when completed and switched on, a machine of this kind will be comparable in some way to a newborn infant. Now the fact of the matter is that we do not take a human baby out of the crib and start teaching it to play chess, manipulate building blocks or speak English. We accept that before any of these things are feasible the child is going to have to 'grow up' quite a lot – and the essence of growing up is the gradual and progressive gathering of experience, or, coming back to the case of a neural net, the construction of a larger and larger state structure.

Only when a child has had two or more years of gathering experience more or less under his own steam and at his own pace do his parents begin the process of a formal education. And even at this stage, we take great care to relate the algorithms that have to be learned, the alphabet, the days of the week, a code of behaviour, to the child's infrastructure of experience. To take the instance of object recognition we have examined in the context of both top-down and bottom-up methods, we do not expect an eight year old to master geometry in the abstract. When the teacher talks about cubes or pyramids, and explains their properties (the sort of rules an AI program uses to recognize such things), he starts from the assumption that his pupils know what a cube or a pyramid is, even if they do not know the proper technical label for it.

Of course, as the educational process advances the pupil may well be required to learn about shapes he has not previously encountered, or not knowingly so – we do not routinely note the fact that what we are looking at is a dodecahedron. But it does seem that in our case we can only assimilate and make full use of formal algorithms if we have previously gathered a good deal of practical, experience based knowledge. Thus it is demonstrable that mastering the discipline of geometry allows us to understand things about cubes which we did not understand before and that this understanding can be extrapolated in order to classify and gain further understanding of more complex shapes. But, as the problems met by AI show, it does not seem that the process can be short-circuited by programming the algorithms and ignoring the importance of experience.

It is worth noting in passing that one of the consequences of the liberalization of educational methods in our schools has been to emphasize the importance of allowing children to discover things for themselves – by self-expression in words or pictures, for instance – at the expense of the more traditional methods of learning by rote. One could, for example, contrast modern methods of language teaching, involving conversation and free discussion, with the traditional classical curriculum which required pupils to learn to parse Latin sentences in a manner very similar to that of the parsing program we looked at in Chapter Nine.

None of this is to say that the computer, preprogramming or AI have nothing to offer the practical science of robotics, or that we should set them aside until the point is reached at which neural nets have learned enough about the world to benefit from a formal education. Quite the reverse, in fact. For, if we are right in thinking that the bottom-up approach may one day culminate in a full scale replica of a brain, then all we will have achieved in utilitarian terms is to produce at enormous costs in time, money and ingenuity, a single example of something we already possess in profusion. It is much more likely that the fact that the digital computer is a different kind of brain is exactly what will continue to make it useful in robotics and elsewhere. There will certainly be applications in which the properties of nets are more useful than those of computers, vision being but one obvious example. But there are also applications in which the computer may continue to reign supreme. It is likely, for example, that it is actually a better kind of brain for controlling very precise muscular movement than our own. Robotics will doubtless find many ways in which the two systems, with their very different characteristics, can profitably work together.

By the year 2000, after another decade and a half of development, *robot industrialis* will have beome an extremely common and diverse object. It is almost impossible to forecast all the shapes it will take and the uses that will be found for it. There seems little doubt that silicon technology will continue to make more and more capable elements for information machines available at lower and lower prices, or that the economic imperatives that gave rise to automation will continue to impel us towards a greater and more widespread use of machines that replace men. What the social and economic costs will be, only time can tell. But Norbert Wiener's warning that the invention of information processing machinery, which supplanted the skills of the mind as the machines of the first industrial revolution supplanted the skills of the hand, posed a threat which could only be averted if we adopted a system of values 'based on something other than buying and selling' looks more than ever relevant.

In the concluding chapter of this book, however, we propose to concentrate on the prospects for *robot fundamentalis* at the end of the millennium, and in particular to try to see how far it might help us to resolve some of the fundamental questions about ourselves which we still find puzzling.

12. The Millennial Machine

The proper study of mankind is man.

Alexander Pope, 'Essay on Man'

We began this book by defining a robot as a machine made in the image of man. Although we have identified many serious problems that will stand between robotics and the realization of this objective, we hope that we have also persuaded the reader that the essentials of such a thing are now beginning to come within our technological grasp. It seems conceivable that sometime in the future, perhaps by the year 2000 or so, roboticists may have reached a point at which they can contemplate the design of an electronic machine which will be, in terms of its capacity, structure and complexity, comparable with our own brain. In this concluding chapter we shall, therefore, examine the implications of such a 'millennial robot' and try to discern its likely nature.

The actual physical form taken by the machine will be of secondary importance, for the image it will be designed to capture is not that of man the biological machine, but man the thinking machine – *Homo sapiens*. In fact, although we will, for convenience, speak of it as a single entity, it may well consist of a whole series of smaller machines each reflecting one facet of the whole: it will even have value if it remains in the realms of the imagination as a coherent body of knowledge and experience, for the specifications for the machine will amount to an understanding of man.

It may well seem that enlarging our understanding of ourselves is a purpose sufficient in itself to justify the cost, in time, trouble and money, of a millennial machine. But it should also be noted that it would serve some extremely important and very practical purposes: for it will provide a new tool for investigating a subject which has proved notably resistant to all the more established and orthodox scientific techniques. The nature of the mind – where it comes from, how it relates to the body, the ways in which it grows and the reasons why it goes wrong – is a subject of the most pressing concern to disciplines ranging from philosophy to medicine, from education to psychology. Yet, in every instance, our ability to think clearly about the problems is limited by the elusive and intractable nature of the mind itself. Despite all our efforts, the ghost in our own machine remains tantalizingly spectral.

The overriding difficulty is our inability to distance ourselves from the thing that we wish to analyse. Since all the evidence we have *about* minds is both collected *from* minds and interpreted *by* minds, objectivity is impossible to achieve. No hypothesis about the mind can be anything other than supposition so long as it is based on data which must be suspect. Furthermore, it can never be fully tested by repeatable experiment so long as the only experimental subject available is another human mind – for very strong and proper reasons we set firm bounds to the use of our own and other people's minds for investigative purposes.

The millennial robot will allow us to overcome this difficulty, for the whole project would clearly be pointless unless it were possible to feel confident that, when complete, the machine would have a property which was at least analogous to what we call a mind. Whether the word 'mind' was used to describe it would be a matter of choice; the point is that unless the machine contained, or could acquire, a ghost it would remain a meaningless jumble of hardware. We have shown that the concept of a state structure allows us to think about what happens inside automata in a way that would be otherwise impossible, and whether we choose to distance ourselves from a mechanical brain by saying that while we have minds it has a state structure seems beside the point. The fact is that the language of automata theory enables us to think about all automata, whether natural or artificial, in a way that is rigorous, logical and corresponds with the realities that we observe.

As has been made clear, we think it highly improbable that when the last bolt has been tightened and the last circuit soldered into place a roboticist will be able to bring a millennial machine 'to life' by inserting a ready made mind into its electronic brain. The experience of those who have tried to devise pre-packaged minds, in the form of AI programs, suggests that a learned state structure not only corresponds more closely with our perception of the natural mind but is also the only kind which will be able to cope with the combinatorial complexities of the real world. We therefore assume that the millennial robot will be based on neural net structures of the kind that have been described in the last two chapters and that among its properties will be those of a learning automaton.

It must be added that it will not be sufficient to construct a machine which comes into existence in a totally blank, unprogrammed form. Even if such a machine would work, it would certainly have departed from the natural model. For it is clear that human babies are born with some innate or inherited mind, enough at least to organize their basic functions and to enable them to obtain food. Many would argue that the degree of preprogramming is much greater than this, and in the case of some animals which can stand up, move about and react to danger within minutes of birth this is clearly so. The question of how far an artificial brain would have to be preprogrammed, and in which ways, is perhaps the central one, and we shall return to it shortly. In the meantime, the fact that at least some proportion of the robot's state structure will have to be learned from its experiences suggests many of the ways in which it might prove valuable.

Many of the outstanding questions about our own minds centre on what happens to them after we are born rather than on how large a proportion of them we are born with. Though the Chomsky-Piaget controversy may turn upon the relative contributions of nature and nurture, education and cognitive psychology pose many other problems to which the example of a learning automaton will clearly be germane. The fact that the machine will have to find out about the world for itself and literally 'make up its own mind' in response to its experience just as a human child does suggests a myriad of murky issues upon which light might be thrown if we were able to keep track of what was going on inside it.

If it is to be useful, the robot will have to accumulate experience that is comparable to that of a human being. It will, therefore, certainly have to be equipped with a set of senses. Sight, hearing and touch would seem to be a minimum sensory repertoire, though it is intriguing to speculate how rapidly and in what ways the state structure of a machine might diverge from the 'norm' were it endowed with senses that are absent or undeveloped in human beings. Would it be the case, for example, that a robot with an extremely sophisticated sense of smell could conjure up 'internal smells' just as WISARD can summon up an 'internal image'?

But senses by themselves will be insufficient. If a machine remains a mere passive receptacle for information its picture of the world will be that of an environment which behaves arbitrarily and which cannot be altered or influenced. Rather clearly, this would lead it to develop a mind very different from our own which is ruled, often dangerously so, by the notion that we can and will affect the state of the world. A body will therefore be required which, although it need not resemble the human physique, must be so designed that it allows the robot to gather the same kind of information that we ourselves collect. It will have to be able to push, pull, poke and probe, pick things up and move about under its own steam.

It is worth noticing that the fact that a body is a pre-eminent requirement of any robot which is going to learn about the world for itself points to the improbability of those immensely superior disembodied minds of which science fiction is so fond. Rather than having the ability to manipulate matter at will, it seems more likely that such creatures would be pathetically ignorant of the kind of reality that we comprehend so effortlessly. But the fact that we invent and entertain such fantasies points towards the confusion we still suffer from when we try to think about the mind.

On the one hand, the sense that a mind is, like a state structure, an incorporeal entity leads to the belief that such a thing might be found divorced from any physical incarnation, a notion which, reduced to its logical conclusion, amounts to the suggestion that the state of a machine can change even if the machine does not exist. On the other hand, the fact that the mind is a property of the brain leads us to become confused when we try to disentangle the one from the other. Having accepted that the brain is susceptible to mechanical explanation at some level, it is difficult to resist the conclusion that it must therefore be a machine in the common sense of the word, and the idea that a machine can contain something that is in it but not of it is disconcertingly unfamiliar. One of the results of this lack of clarity has been that, when dealing with mental illness for example, it has often been assumed that because the mind is in the brain, and the brain is an organ of the body like the heart or the liver, the problems of the mind could somehow be resolved by 'treatment', as one sets a broken limb or treats an infected organ. In retrospect, the belief that a lobotomy might cure a mind looks absurd; one might as well try to correct a computer program with a hacksaw. But the belief that minds are amenable to surgical tinkering has taken a long time to die.

The fact that one can consider the relationship between the electronic hardware of a robot and the state structures it incorporates in the same way as that between a brain and the mind it embodies may help us to clarify our thinking about what goes wrong with minds and why. For example, psychologists now generally agree that many forms of mental illness are in part a consequence of experiences which occurred in the past, of the kind of mind that got into the brain. The existence of a machine with state structures that could be created and manipulated at will and without compunction could be of inestimable importance. This is not to say that a 'clinical robot' would resolve all the issues that puzzle psychology, many of which may have their roots in chemical and biological causes for which an electronic machine built of silicon chips can provide no analogies, but it will at least provide a framework of comparison which we so far lack.

In Chapter Nine, for example, we mentioned the 'paranoid' computer program, PARRY, devised by Mark Colby. This, it seems to us, is essentially a description of paranoia rather than an explanation. The program's behaviour is certainly very similar to that of a person who has become obsessed with the idea that he is being persecuted by the Mafia, and it throws some light on the sort of 'script using' mechanisms that might be operating in such a person's mind, but it does not help us to understand how or why such mechanisms might become lodged in a mind. If we accept, at least for the sake of argument, that an individual suffering from paranoia is behaving in ways that are perfectly consistent with his own internal logic but based on profound misapprehensions about the way in which other people actually behave, then we can see ways in which a clinical robot might help to determine the cause of the problem. In terms of automata theory, one could say that at some point in the past that part of a paranoid person's state structure which represents his world automaton has gone awry with the result that any input causes the same output, whether it is appropriate or not. A machine which enabled psychologists to investigate how and why state structures become malformed in this way might well have something to contribute towards the analysis of the condition, and help for the afflicted.

If we are prepared to treat our bodies with drugs on the basis of the results they produce in animals, then it seems perfectly reasonable that we might experiment with and assess methods for treating the mind by trying them out on machines. To adopt this expedient need not involve the conclusion that machines are our equals, only the assumption that they can be similar in *some* respects. In one direction at least, it already seems clear that the comparison will be extremely reassuring to our sense of our own dignity.

Even in the case of WISARD, a machine with a capacity that is but a tiny fraction of that which a millennial robot would have, it has become obvious that state structures which are far from predictable or determined can come into existence. Given that the state structure of a learning automaton is a product of the history of its inputs, and that, if it is to work effectively, those inputs will have been linked together by complex chains of association, one can only know what the state structure of such a machine is if one has a full and complete record of all its inputs and the sequence in which they occurred. In laboratory conditions, such a thing may be possible, but it is of course axiomatic that laboratory conditions would not, by themselves, suffice to provide the millennial robot with all the experience it would require if it was to develop a state structure on the scale of a human mind.

In practice, while it will obviously be essential to keep track as closely as possible of all the sources of information which are contributing to the machine's expanding state structure, it seems likely that no system of record keeping will comprehend all of them. In small ways, if not in large ones, the mind of the machine will quickly expand to the point at which it is both unique and unknowable.

Nor will it be feasible to keep track of the state structure at the basic mechanical level of the settings of the components of the neural nets. As we have seen, the effectiveness of a net depends crucially upon its capacity for generalization and this means that it is impossible to attribute a particular internal state to a particular input with any degree of precision. One cannot, for example, say that in the case of WISARD the RAMs were set in a particular way by one particular image and that, therefore, that image and only that image will obtain a response from the net. The sheer impracticality of making this kind of prediction can be illustrated by the fact that it would take one of the world's largest and fastest computers three hours to perform the calculations needed to simulate the recognition of just one of the patterns that can be formed by the eight million bits in one of WISARD's discriminators – a feat that the machine itself performs effortlessly in a matter of moments.

Even if we were able, by some miraculous means, to keep track of all the settings of all the RAMs in a brain with a capacity of billions of bits, the fact would still remain that, in the case of a human brain, the nerves which transmit the inputs to the neural synapses are (indeed, as we have seen, must be) connected up at random. This in itself, incidentally, makes nonsense of another science fiction favourite, the brain transfer machine which extracts the contents of one brain and inserts them in another. For the fact that an external input has been translated into an internal state by a random network of nerves guarantees that patterns which were significant to the brain which created them would be meaningless to another brain.

In the case of robots, however, having devised one system of random connections it would be possible to duplicate it, and this raises the interesting possibility that, once the expenditure of effort and time necessary to educate one robot has been made, one could clone it by transferring its entire state structure to other robots. It does not follow that in the process of cloning one would discover the significance of the patterns that were being transferred: the minds of all the machines would remain unknown and unknowable in any detail, but they would, at the moment of transfer at least, be identical.

Two major problems will have to be solved before the millennial robot becomes a practical possibility. The first concerns the way in which the brain itself is organized. We know enough to understand that it is not simply one vast undifferentiated network of neurons, but that, on the contrary, it is made up of a very large number of separate nets and that the precise relationship between them is crucial to its functioning. But the role played by each network and the influence it exerts on the others still has to be established.

This is a puzzle which may well be solved by collaboration among neurophysiology, psychology and robotics. Physical examination of the brain, coupled with observation of its operation, allows researchers to make educated guesses about what is going on in different areas of it, and robotics can offer electronic testbeds on which those hypotheses can be put to the test. It may be, for example, that a manipulator coupled to a system of neural nets would provide a means of investigating further the way in which whole sequences of muscular movement, once learned, are filed away in the cerebellum for future use. The work of David Marr, which we mentioned in Chapter Eight, has indicated that this is how the brain copes with physical functions which have to be learned but which, once mastered, are repeated without apparent effort. But it is an open question whether patterns are stored in the cerebellum simply as a result of repetition, as the result of some decision reached elsewhere in the brain, or when they have become sufficiently generalized and clarified to be useful.

The fact that many physical skills like swimming or bicycle riding are mastered suddenly when, after many fruitless attempts, we 'get the hang of it' suggests that some process of refinement is taking place and that it is only when inputs and feedback are resolved into a satisfactory output pattern that the internal state is transferred to permanent storage. On the other hand, the fact that, once acquired, even unwanted patterns of movement like a faulty golf swing are difficult to eradicate indicates that once set to deliver a particular response the nets within the cerebellum can be adjusted only with difficulty.

The task of disentangling the hierarchy of automata that makes up the brain is one that seems likely to be achieved only by painstaking investigation. We can feel reasonably confident that the answers will be found, but exactly when and how it is impossible to estimate. The second problem looks much less amenable to this sort of treatment and may require a bold leap into the dark. It centres, of course, on the question raised earlier in the chapter, of how much preprogramming a millennial robot would require in order to 'get it off the ground'.

It must be assumed that a high proportion of the preprogramming in the human brain is concerned with functions which would have no counterpart in the millennial machine. It will not have a heartbeat or a metabolism to regulate and it will be assured of a steady supply of energy. Some time in the future it may be advantageous to explore the possibilities of a self-sustaining robot which forages for its own living, but in the first instance at least it seems unlikely that the millennial robot will be given responsibility for the routine administrative chores involved in keeping itself ticking over. It also seems unlikely that it will be equipped with appetites.

This is a contentious issue, since it is quite possible to argue that all the drive and curiosity which lead an animal to expand its experience of the world originates in the need to satisfy its hunger and avoid pain. On this basis a brain which was not preprogrammed with the imperatives required to obtain food and ensure its own survival would lack the catalyst that gives rise to a mind. In the case of human beings, however, we must suppose that something more is involved. If our minds grew only in response to our needs as biological structures then it would be hard to account for a whole range of human activities which seem to confer no practical advantages upon us but which, we instinctively feel, are precisely what separates us from 'lower' animals. In particular, as scientists themselves, roboticists might be concerned as to why we are so curious about nature itself, including our own. The question might be put succinctly by asking: what kind of machine is it that wants not only to feed but also to understand itself?

The search for the appropriate grain of sand which must be inserted into the electronic oyster to ensure that it grows a mind seems to lead back to the issue of intentionality which we discussed in Chapter Nine. This is already becoming clear in the case of WISARD as it reaches its third stage of development. This, it will be recollected, involved equipping it with a window of high definition which it could move about in order to concentrate upon certain areas or features within its overall field of view, thus simulating the mechanisms that enable human beings to focus their attention on that which they find interesting or significant. As described, the experiment required the machine to be taught to move the window in certain prescribed patterns or in response to certain kinds of images. While this certainly reproduces some features of human behaviour – that of a student who is told to 'look at that' or 'concentrate on this' – the fact is that for most of the time we devote our attention to what we ourselves find interesting.

What kind of mind would WISARD, or a more ambitious machine, have to be given in order to ensure that it too 'wanted' to find things out and did not have to be taught all that we wished it to know? In programming terms, and programming of some kind is after all what we are talking about, the need is for some algorithm that encompasses the motives that underlie intelligence. To discover such a thing we might look in two directions, firstly at AI, the field which concerns itself with devising algorithms for machines, and secondly to psychology which tries to discover the algorithms that direct our own behaviour. It would obviously be especially significant if some common ground could be found between the two, some algorithm devised by AI for which psychology has detected a natural counterpart.

Unfortunately, AI seems unlikely to provide much help in this quest; indeed, many of its leading theorists would question the whole undertaking. The first difficulty is that there is a rift within the field itself between those who believe that AI programs are, at some level, a valid model of human intelligence and those who argue that their task is simply to find ways of making computers behave as if they were intelligent without worrying about whether the means are comparable with those that our own brains employ. The first point of view is put by the British philosopher Margaret Boden when, in her book *Artificial Intelligence and Natural Man*, she defines AI as 'research that is somehow relevant to human knowledge and psychology'. The opposing case is represented by the American Nils Nielsson, who suggests that he and his colleagues would do well, for the time being at least, to 'limit our attention to those principles that are relevant to the engineering goal of building intelligent machines'.

Even if Boden's argument is accepted, the problem remains that AI has not only failed, as John Searle points out, to encompass intentionality, but has also deliberately eschewed it. The response of those who support 'hard AI', led by the philosopher Daniel Dennett, to the criticisms levelled by Searle (which are discussed in Chapter Nine) has been to dispute the basis of Searle's case. The fact that the programs Searle criticizes seem to suffer from an absence of meaning or intentionality is, Dennett says, not a consequence of some inherent failing but just a reflection of their comparative simplicity. Once AI reaches the stage at which programs can cope with really complex tasks it will become irrelevant whether they have any built-in reason for performing them; to the observer they will appear to have such a reason.

The essence of Dennett's argument can be illustrated by taking the example of a chess playing machine. The programmer may be aware that for 90 per cent of the time the machine has no idea that it is trying to achieve checkmate and is merely obeying the dictates of its heuristics, but to an opponent it *seems* that the machine is trying to win the game. If a machine appears, in practice, to have a purpose, is there any point, Dennett asks, in worrying about whether it has one in principle?

From the point of view of those concerned with robotics as offering a serious model of man this argument seems doubly disingenuous. Since a sense of purpose and meaning is something of which we are extremely conscious in ourselves it seems unnecessarily dismissive to treat it as an illusion, especially since it seems to provide at least a partial explanation of our ability to perform some functions with which AI has rather notably failed to cope. Even more important, perhaps, it can lead to rather serious misapprehensions about machines. This point was made by another researcher, John Andreae, in the late 1960s.

Andreae constructed a simple machine consisting of a box on which were mounted three push buttons and a single light. The functioning of the machine was outwardly mysterious; sometimes the light came on when a button was pressed and sometimes it did not. Andreae then invited a number of subjects to experiment with the box and try to explain the rationale that lay behind its behaviour. In fact, the machine was entirely probabilistic: there was one chance in four of the light coming on when one button was pressed, a one in two chance with another button and three chances in four with the third. But only the most perservering of the subjects guessed the correct answer. Most preferred to construct highly elaborate theories about the sequences in which buttons were pressed and others simply gave up and credited the machine with 'a will of its own'.

Rather than supporting the belief that, if it seems to have intentions, one might as well treat a machine as if they existed, this experiment seems to suggest that our own sense of meaning and purpose is so strong that we are reluctant to believe that even inanimate objects can be without one.

If we accept that AI either cannot or will not provide the kind of algorithm we require and turn instead to psychology we are immediately faced with a wealth of hypotheses and conjecture. However, the field can be narrowed down considerably in the light of three factors. First, the vast bulk of psychological experiment and theory has been concerned with what happens to the mind in the course of its development rather than with its innate, aboriginal nature. Second, it seems clear that the value of the millennial machine will be enhanced if its preprogramming is as economical as possible. If the kind of mind it develops is predetermined by the kind of mind it is given, then the chances of our learning something interesting are much diminished. It might, for example, be feasible to produce a program which incorporated the elements of Freud's theories about the nature of the mind – but to do this would only ensure that any machine which made use of it would behave in a Freudian fashion. It would obviously be desirable to restrict as far as possible the extent to which programming predetermines the outcome of the whole experiment. Finally, of course, some schools of psychology are eliminated simply because their ideas seem to be incompatible with the kind of brain which we assume the millennial robot will have.

In our view, of all the relatively well-established and respected psychologies, the one which best fits the bill is the theory of personal constructs developed by the American psychologist George Kelly in the 1950s. As set out in full in his book, *The Psychology of Personal Constructs*, Kelly's arguments occupy three full volumes and the subsequent literature amounts to over three hundred papers of one kind or another. However, one of the great advantages of the theory, from the roboticist's point of view, is that Kelly was deeply concerned with the kind of rigour and precision that is more often found in mathematics than in psychology. He therefore set his ideas out formally as a fundamental theorem with eleven corollaries, and it is possible to discuss how these might translate into the elements of a program without doing too great an injustice to the subtlety and complexity of his case.

A program of research has been underway for some time at Brunel University aimed at identifying and clarifying the essentials of personal construct theory and reformulating them in a shape which could be used to either write an orthodox computer program or provide a basic program for a machine based on neural nets. Any reader wishing to pursue the subject in greater detail will find the translations of the fundamental axiom and all eleven corollaries in

Appendix II; for our present purposes we will concentrate on the fundamental axiom and those corollaries which seem most immediately relevant to a millennial robot.

The fundamental axiom, Kelly's summary of what he sees as the crux of the matter, reads as follows: 'A person's processes are psychologically channellized by ways in which he anticipates events.' Since this statement is a highly condensed summary of an elaborate idea, some gloss is required before we suggest how it could be 'translated' into a prescription for a robot mind. 'A person's processes' means, broadly speaking, what goes on inside someone, their entire inner workings. 'Psychologically channellized' means 'given an information structure', almost 'made sense of', and the phrase can be translated without doing any violence to Kelly's meaning as 'turned into a state structure'. 'By ways' seems self-explanatory, but it is important to note the use of the plural which suggests that there is more than one way in which a state structure can emerge. By 'anticipates events', Kelly's own arguments make clear, he means 'predicts new inputs by reference to previous ones'.

The whole axiom can, therefore, be translated into terms which are applicable to robotics in the following form: 'What goes on inside a robot (and the assumption is, of course, that we are talking about a robot which is a valid model of man) is turned into a state structure by ways in which it predicts new inputs by reference to previous ones.' Encouragingly, this could serve, without any amendment whatsoever, as a definition of the kind of learning automata we discussed in Chapter Seven or for a neural net with feedback. For example, when WISARD learned to associate the image of John with the sequence of J-O-H-N it used its internal electronic processes to create a state structure which predicted a new input, the picture of John, whenever it saw a past input, the sequence of letters, or vice versa.

So far, it seems, so good. Psychology and robotics are talking about the same kind of mind. Kelly's first corollary takes things a stage further. It reads: 'A person anticipates events by construing their replication.' By construing, the original text makes clear, Kelly meant 'placing an interpretation upon', or 'attaching a label to'. In the crudest terms, the corollary means that because the sun has always risen every morning we expect the event we call dawn to occur tomorrow. Again this is entirely consistent with the behaviour of a learning automaton and the capacity for generalization enjoyed by neural nets allows it to attach a common label to a whole group of inputs even if they are not precisely the same. Thus WISARD expects J-O-H-N when it sees John because the two have previously appeared together, and its idea of John is sufficiently general for it not to matter if his expression changes.

At the heart of the theory is the idea of what Kelly calls the 'dichotomous construct'. Essentially, the suggestion is that the mind consists of a large (though always finite) number of constructs or, less precisely but perhaps more helpfully, subject areas. The range of constructs in a mind increases with experience and every input is assigned either to an existing construct or becomes the basis for a new one. Furthermore, each construct takes the form of a scale defined by two extremes. An example may make things clearer. As a person's reading skills develop they will generate a whole series of constructs such as: good plot/bad plot; well-written/badly-written; exciting/dull; fiction/non-fiction . . . and so forth.

The interpretive or labelling mechanism assigns an input to the appropriate construct on the basis of past experience and also ranks it somewhere on the scale attached to that construct. The mind organizes its experience in terms of the relationship between different constructs in a fashion which is characteristic of each individual and which it finds most useful in anticipating future events.

Although this sounds complicated it is relatively easy to envisage the mechanism in terms of state structures. We have already seen how the reappearance of a familiar input will drive an automaton into a particular 'learned' state: this is the equivalent of an input being assigned to a construct. Since it is also a property of automata that one state leads to another, according to the dictates of 'if . . . then . . .' instructions which may be indeterminate (that is, State A may lead *either* to State B *or* State C), a branching state structure, in which the value attached to the input determines which branch will be followed, would achieve the desired result.

We need not, however, concern ourselves too much with precisely how an assembly of neural nets would achieve the requisite state structure. The crucial point, for our purposes, comes with Kelly's 'Choice Corollary'. This states that: 'A person chooses for himself that alternative in a dichotomized construct through which he anticipates the greater possibility for extension and definition of his system.' This implies that one of the basic constructs in the mind, perhaps the fundamental one that is necessary to create all the others, is a growth/no growth construct and that this determines that, in deciding where to rank an input on the scale of all relevant constructs, the mind will always choose the course that seems likely to lead to its own expansion.

Kelly is thus arguing that the mind (or at least that part of it which most concerns him and is of most interest to us, the thinking, intelligent part) is driven by curiosity pure and simple. Translated into terms which can be applied to the millennial robot, this means that all we have to do to ensure that its mind develops to the fullest possible extent, while still remaining comparable with a human mind, is to equip it with a single preprogrammed imperative: enlarge your own state structure in ways which best enable you to predict and control the world.

A built-in instruction of this kind would certainly be sufficient, for example, to ensure that WISARD did not leave its window of attention stationary and purposeless. It would be driven to move it around an image and to learn everything that it could from that image. Equipped with this single overriding need to organize all experience in ways best designed to gain further experience it would indeed seem that a millennial robot would incorporate a recognizable intentionality and that, like human beings, it would have insatiable appetite for giving meaning to the world in which it found itself.

Whether a robot psychology based on personal construct theory will result in a machine which reflects a recognizable image of ourselves we shall only discover when we are in a position to put the proposal to the test. But before abandoning the subject it is worth noticing the implications of two of Kelly's other corollaries.

The first of these, the 'Commonality Corollary', is the one which has allowed personal construct theory to be put to practical use by psychologists. It states that: 'To the extent that a person employs a construction of experience which is similar to that employed by another his processes are similar to those of another person.' By use of what is known as a Repertory Grid, psychologists have been able to elicit the 'constructions of experience' employed by other people (how this is done is outside our scope, but it will suffice to say that psychology has found such grids to be useful and effective tools). From the roboticist's viewpoint the importance of the corollary is that it suggests a means by which we might understand, even empathize with, what was going on 'inside the head' of a millennial robot without relying upon the strictly mechanical language of electronic engineering or the purely descriptive system of automata theory – we could, in short, understand the robot *as if* it was another person.

This notion is taken a stage further by the 'Sociality Corollary': 'To the extent that one person construes the construction processes of another he may play a role in a social process involving the other person.' If this holds true for robots as well as people, then it suggests that, at the intellectual level at least, the relationship between the millennial machine and its makers may go beyond that of experiment and experimentalist. Not only will roboticists be able to understand their robots, robots may begin to understand roboticists.

We have ended this book on a note of speculation – many might say a flight of fantasy – because it seems to us that roboticists have nothing to lose, and the rest of us may have much to gain, if the idea of a machine made in the image of man is taken seriously. At some point this will require those involved in the science of robotics to think and act boldly. There is no reason to think that this will lead the discipline to abandon either responsibility or humility. Those whose job it is to try and build robots have better cause than most to appreciate that the image they are trying to capture is one of almost infinite complexity and intricacy.

What have we to fear if a millennial robot should at some point see the light of day? No more, we suggest, than we need fear from the birth of a child. It will be, like every other machine, one of the 'children of our brain' and exceptional only in that it will offer us the chance of achieving an end from which nothing but good can flow: a fuller understanding of ourselves.

Appendices

Appendix I: A Brief History of the WISARD Project

The functional similarity between a neuron and a RAM element was first recognized by one of the authors (IA), then working at Queen Mary College in London, in the mid-1960s. At that time RAMs were still rare and expensive pieces of equipment. Computers relied upon magnetic core structures for the vast bulk of their memory and the RAM capacity of even quite a large machine might consist of only one element with 32 address terminals or a storage capacity of 2^{32} bits. Although the idea that it might be possible to build a device containing thousands of elements seemed impossibly far fetched, it was becoming evident that the silicon chip technology which had already given birth to the first microprocessor would, one day, be applied to memory elements. In the meantime, experience could be accumulated either by constructing very simple hardware versions or by computer simulation.

The team at Queen Mary College chose the former alternative and, while waiting for silicon technology to catch up with their ambitions, set about designing a substitute for the RAM which would be feasible and economic to manufacture in reasonable quantities. The result was the SLAM (Stored Logic Adaptive Microcircuit) which was produced in collaboration with, first, the Plessey Company, and later, Integrated Photomatrix Ltd. The SLAM's firing characteristics are identical with those of the RAM, and the early models built by Plessey were indeed the first silicon memory elements produced anywhere in Britain.

By 1969, when the first three-address SLAMs became available, the team had moved to the University of Kent, where they constructed the first flexible neural net that had the potential to expand into a larger system (other contemporary devices developed by Bernard Widrow in the US and W K Taylor in Britain were, as described in Chapter Ten, reliant upon elaborate mechanical systems). The machine, called SOPHIA, used twelve SLAMs to cover a 6×6 matrix and had only one discriminator. Though very valuable for the experience it provided of the engineering and electronic problems involved in building nets of this kind, SOPHIA was obviously of little practical use.

By this stage, the silicon RAM was a reality. Sixteen-bit chips were being manufactured in the US, but were still very expensive and hard to come by. The team at the University of Kent therefore stuck to SLAMs for its next venture which was considerably more ambitious and was funded by the Science Research Council. The new machine, completed in 1972 and dubbed MINERVA, contained 4096 four-input SLAMs arranged in 16 separate discriminators to cover a 16 × 16 matrix. It proved highly successful, coping with recognition problems such as those involved in handwritten characters and offering a speed of response which was far better than anything conventional computers could manage.

The next development took place at Brunel University, to which IA and his colleagues had by now moved, and took advantage of the enormous boom in minicomputers which occurred in the early 1970s. This time their efforts went, not into a machine, but a piece of computer software which simulated the behaviour of a net. The program was called JANSYS (for J A Napper System, after its designer) and allowed the team to experiment with 16 × 16 windows in a 64 × 64 matrix – the origins of WISARD's third stage of development described in Chapter Eleven. JANSYS also allowed the team to do the groundwork necessary before undertaking the construction of a really large net able to deal with high quality images on a TV-sized 512 × 512 matrix.

By 1979 RAM elements could be bought off the shelf and the time was ripe for work on WISARD itself, in its original form, to begin. Bruce Wilkie joined the team on an SRC grant to engineer the giant machine, which was completed in June 1981. It has now been patented and is available in several different versions as an all-purpose pattern recognition device. In the meantime, the original machine has begun to undergo the program of expansion and refinement described in Chapter Eleven.

Appendix II: Outline for an Intelligent Program

This appendix sets out the fundamental axiom and the eleven corollaries which make up G A Kelly's theory of personal constructs, together with 'translations' of each statement into a form which could be incorporated into a computer program. It should be noted that although the translations are, in their present form, designed specifically as the basis for an orthodox computer program and this accounts for the divergence from the more general version of the fundamental axiom given in Chapter Twelve, there is nothing to suggest that only a von Neumann computer could make use of such a thing. It is perfectly possible to re-translate each statement into a form which would be appropriate to a neural net structure by substituting the words 'robot mind' for 'intelligent program'.

The Fundamental Axiom

'A person's processes are psychologically channellized by ways in which he anticipates events.'

The translation into a computational form starts with a program which may be called 'The Intelligent Program (or IP)'. This is the translation of 'a person's processes'. The complete list of such translations is given below followed by the translation of the entire axiom.

Kelly	*Algorithm*
Person's processes	Intelligent Program (IP)
Psychologically	Defines the range of convenience of the IP to 'thoughts and actions' as being the state labels of the IP program (as opposed to, say, digestion or skin texture).
Channellized	The IP has a structure (flow chart or state transition diagram or a tree, etc).
Ways	Points to a possible non-determinism in the structure.

Anticipates	States in the IP structure are related to previously experienced inputs.
Events	Input to the IP related to world realities.

Intelligent Program Fundamental Axiom

An Intelligent Program *about* human thought and action has a non-deterministic structure that is predictive of its own input.

Kelly's Construction Corollary

'A person anticipates events by construing their replications.'

IP Construction Corollary

The IP *becomes* predictive of input by assigning interpretation labels to similarity groupings in the input stream.

Kelly's Individuality Corollary

'Persons differ from each other in their construction of events.'

IP Individuality Corollary

Different instantiations of the IP may assign interpretation labels to input streams in different ways.

Kelly's Dichotomy Corollary

'A person's construction system is composed of a finite number of dichotomous constructs.'

IP Dichotomy Corollary

Interpretation labels in the IP are called constructs. They are organized as an associative data base so that input-related elements address a construct and obtain an applicability value between two further elements that make up the construct.

Kelly's Organization Corollary

'Each person characteristically evolves, for his convenience in anticipating events, a construction system embracing ordinal relationships between constructs.'

IP Organization Corollary

IPs will update and structure the construct data base so as to include non-symmetric relations between constructs and thus aid the anticipation of input.

Kelly's Choice Corollary

'A person chooses for himself that alternative in a dichotomized construct through which he anticipates the greater possibility for extension and definition of his system.'

IP Choice Corollary

New elements are assigned by the IP to constructs so as to maximize the potential growth of the data base.

Kelly's Range Corollary

'A construct is convenient for the anticipation of a finite range of events only.'

IP Range Corollary

A construct in the IP data base may be addressed by a finite range of elements only.

Kelly's Experience Corollary

'A person's construction system varies as he successively construes the replication of events.'

IP Experience Corollary

The IP varies its construct data base step by step, in response to its ability to interpret and predict events.

Kelly's Modulation Corollary

'The variation in a person's construction system is limited by the permeability of the constructs within whose ranges of convenience the variants lie.'

IP Modulation Corollary

The amount by which the IP can vary its construct data base is limited by size of the element sets addressing each construct.

Kelly's Fragmentation Corollary

'A person may successively employ a variety of construction subsystems which are internally incompatible with each other.'

IP Fragmentation Corollary

The IP may address, in turn, parts of its construct data base that are logically incompatible with each other.

Kelly's Commonality Corollary

'To the extent that a person employs a construction of experience which is similar to that employed by another, his processes are similar to those of another person.'

IP Commonality Corollary

To the extent that a person employs a construction of experience which is similar to that employed by the IP (as measured by a repertory grid), his processes are similar to those of the IP.

Kelly's Sociality Corollary

'To the extent that one person construes the construction processes of another he may play a role in a social process involving the other person.'

IP Sociality Corollary

To the extent that the IP construes the construction processes of a person, it may play a role in a social process involving this person.

Further Reading

In a book of this nature, intended primarily for the non-specialist reader, a formal apparatus of notes and references would seem out of place. Many of the principal sources are identified in the text and the reader who is interested in following up the subject, or some aspect of it, will find that the more approachable and accessible texts are listed below. This is by no means an exhaustive bibliography of robotics; such a thing would, inevitably, contain a preponderance of references to scientific papers and learned journals which would, perhaps, tell the general reader rather more than he wanted to know; it does include both those books which we ourselves have found useful in the course of preparing this work and a core of 'essential' texts which are indispensable to anyone who wishes to understand contemporary robotics in any detail.

Chapter One

Many authors have touched upon the role of the robot in literature and a useful collection of articles may be found in Geduld, H M and Gottesman, R, *Robots, Robots, Robots* (New York Graphic Society, Greenwich, Conn, 1978) which also contains a filmography of robots. A fuller and more scholarly examination of the idea of 'unnatural man' and a brief history of automata are provided by Cohen, J, *Human Robots in Myth and Science* (Allen & Unwin, London, 1966).

A complete catalogue of all the fictional treatments of robots would occupy a book in itself, but no list of further reading would be complete without the three seminal works of fiction: Shelley, M, *Frankenstein* (originally published in 1819, most recent edition Oxford University Press, Oxford and New York, 1971); Capek, K, *RUR (Rossum's Universal Robots)* (Doubleday, New York, 1923); and Asimov, I, *I, Robot* (currently available in Panther Books, London, 1968 and Fawcett Books, New York, 1970).

A M Turing's classic paper, 'Computing Machinery and Intelligence', in which the Turing test was first proposed, was originally published in the journal *Mind* and is reprinted in Hofstadter, D and Dennett, D (eds), *The Mind's I* (Harvester Press, Brighton, 1981 and Basic Books, New York, 1980). A full account of the early disputes over the nature and status of computer intelligence, at least from the British point of view, can be found in Hodges, A, *Alan Turing: The Enigma* (Burnett Books, London, 1983 and Simon & Schuster, New York, 1984).

Those interested in a critique of the behaviourist concept of intelligence will find a highly articulate one in Chomsky, N, *Problems of Knowledge and Freedom* (Fontana Paperbacks, London, and Vintage Books, New York, 1972).

Chapter Two

A recent 'classic' which provides a relatively up-to-date outline of industrial robotic practice is Engelberger, J, *Robotics in Practice* (Kogan Page, London, 1981 and Amacom, New York 1983). A good idea of the period of gigantism in computing can be obtained from two books: Turn, R, *Computers in the 1980s* (Columbia University Press, New York, 1974) and Malik, R, *And Tomorrow . . . the World: Inside IBM* (Millington Press, London, 1975). Contemporary industrial robots get at least a mention in nearly all the recent spate of books dealing with the advent and implications of the microprocessor, among the most thorough and readable of which is Large, P, *The Micro Revolution* (Fontana Paperbacks, London, 1980).

Chapter Three

Engelberger (see above, Chapter Two) provides a workmanlike summary of robot anatomy and drives; a more detailed discussion can be found in the 'Robot Technology' series, Coiffet, P and Richard, P J (eds) (eight volumes) (Kogan Page, London, and Prentice-Hall, Englewood Cliffs, NJ, 1983 and in preparation), which offers a full account of robot structure and components. A primer on the subject is also available in Chirouze, M and Coiffet, P, *An Introduction to Robot Technology* (Kogan Page, London, 1983 and McGraw-Hill, New York, 1983).

The Royal Society of Arts in London is shortly publishing an interesting collection of papers on the future of robots.

Those interested in robots with legs will find of interest Raibert, M H and Sutherland, J E, 'Machines that Walk', *Scientific American* (Jan 1983).

Chapter Four

The standard work on the history of the digital computer is Goldstine, H H, *The Computer from Pascal to von Neumann* (Princeton University Press, Princeton, NJ, 1972). Another excellent account is given in Bowden, B V, *Faster than Thought* (Pitman, London, 1953) and a briefer and more popular version is Evans, C, *The Making of the Micro* (Gollancz, London, and Van Nostrand Reinhold, New York, 1981).

Von Neumann's classic paper ('The Report on the EDVAC') was published as 'A Preliminary Discussion of the Logical Design of an Electronic Computing Instrument' (Princeton Institute of Advanced Studies, 1946).

Those interested in pursuing the fascinating story of Babbage and his 'analytical engine' are recommended to read the recent biography Hyman, A, *Charles Babbage: Pioneer of the Computer* (Oxford University Press, Oxford, 1982 and Princeton University Press, Princeton, NJ, 1982).

Hodges (see above, Chapter One) gives a full account of the hitherto neglected contribution of A M Turing to the British computer effort as well as providing a detailed discussion of the ideas which led Turing to the concept of his machine. Further information on the British aspect can be found in Randell, B (ed), *The Origins of Digital Computers* (Springer Verlag, Berlin, 1973).

Chapter Five

The definitive work on cybernetics is still Wiener, N, *Cybernetics* (MIT Press, Cambridge, Mass, 1947), but a briefer summary can be found in Ross Ashby, W, *An Introduction to Cybernetics* (Methuen, London, 1964 and John Wiley, New York, 1963). For Wiener's own misgivings about the possible misuse of his ideas in a social context, see Wiener, N, *The Human Uses of Human Beings* (Houghton Mifflin, Boston, 1950).

Two good introductions to knowledge about the make-up of the natural brain are Eccles, J, *Understanding the Brain* (McGraw-Hill, New York, 1977) and Granit, R, *The Purposive Brain* (MIT Press, Cambridge, Mass, 1977).

Chapter Six

A somewhat starry-eyed vision of the automated future was

presented by Sir Leon Bagrit in his 1964 Reith Lectures on BBC Radio, subsequently published as *The Age of Automation* (Weidenfeld & Nicolson, London, 1965). For the more cautious approach of Wiener see *The Human Uses of Human Beings* (see above, Chapter Five) and Wiener, N, *God and Golem, Inc* (MIT Press, Cambridge, Mass, 1964). Engelberger (see above, Chapter Two) discusses in some detail the application of robots to a whole range of industrial tasks, and Geduld and Gottesman (see above, Chapter One) give plentiful examples of imaginary household robots.

Chapter Seven

The collection of papers which marks the 'birth' of automata theory as a discipline is Shannon, C E and McCarthy, J (eds), 'Automata Studies', *Annals of Mathematical Studies No 34* (Princeton University Press, Princeton, NJ, 1956). A practical approach to automata theory, for those interested in its applications to engineering, is: Aleksander, I and Hanna, F K, *Automata Theory: An Engineering Approach* (E J Arnold, London, 1976 and Crane Russak, New York, 1976).

An account of the debate between Chomsky and Piaget is given in Piatelli-Palmerini, M (ed), *Language and Learning* (Routledge & Kegan Paul, London, and Harvard University Press, Cambridge, Mass, 1980).

Some of the simpler aspects of automata, learning machines and man-machine comparisons are discussed in Aleksander, I, *The Human Machine* (Georgi Publishing, St Saphorin, Switzerland, 1978).

Chapter Eight

For those who wish to know more about artificial intelligence the following two books probably provide the best guides as well as an interesting contrast in emphasis. Boden, M, *Artificial Intelligence and Natural Man* (Harvester Press, Hassocks, Sussex, 1978 and Basic Books, New York, 1978) is much concerned with the philosophical parallels between machine intelligence and human psychology. Nielsson, N J, *Principles of Artificial Intelligence* (Tioga Press, Palo Alto, Cal, 1980) puts the emphasis on the techniques of the business and its applications in engineering.

A valuable, brief introduction to expert systems can be found in Michie, D, 'Expert Systems', *Computer Journal* (Vol 23 No 4, 1980). The logical philosophy of the PROLOG language is well explained in Kowalski, R, *Logic for Problem Solving* (North-Holland, New York, 1979).

Chapter Nine

Boden and Nielsson (see above, Chapter Eight) both cover work in vision as well as other aspects of AI. For a detailed account of the notion of semantic nets and learning systems based on them, see Winston, P H, *Artificial Intelligence* (Addison-Wesley, Menlo Park, Cal, 1977).

Two recent works have explored the whole question of computer understanding of language. They are Winograd, T, *Language as a Cognitive Process* (three vols) (Addison-Wesley, Menlo Park, Cal, 1983) and Wilensky, R, *Planning and Understanding* (Addison-Wesley, Menlo Park, Cal, 1983).

John Searle's article, 'Minds, Brains and Programs', is reprinted in Hofstadter and Dennett (see above, Chapter One) which also discusses the whole issue.

Chapter Ten

The established classic on brain-computer comparisons is von Neumann, J, *The Computer and the Brain* (Yale University Press, New Haven, Conn, 1963). The major work on the behaviour of non-digital neural models and its computer simulation is Rosenblatt, F, *Perceptrons: The Principles of Neurodynamics* (Spartan Books, New York, 1962) and the major critique of this approach is Minsky, M and Papert, S, *Perceptrons: Computational Geometry* (MIT Press, Cambridge, Mass, 1969).

A most readable account of our knowledge of neural systems, especially with reference to vision, is given in Blakemore, C, *The Mechanics of Mind* (Cambridge University Press, Cambridge, 1977).

Technical details of single layer nets of RAMs and their behaviour may be found in Aleksander, I and Stonham, T J, 'A Guide to Pattern Recognition Using Random Access Memories', Institute of Electrical Engineering *Journal of Computers and Digital Technology*, (Vol 2 No 1, Feb 1979).

Chapter Eleven

Feedback strategies as used in WISARD are described in greater detail in Aleksander, I, 'Emergent Intelligence from Pattern Recognition Nets' in *Artificial Vision for Robots* (ed I Aleksander) (Kogan Page, London, 1983).

An early discussion of 'cycles' in brain function may be found in Hebb, D O, *The Organization of Behaviour* (Wiley, New York, 1949). A discussion of the 'state structures' in the brain is contained in Aleksander, I, *The Human Machine* (see above, Chapter Seven).

Chapter Twelve

Apart from Boden and Nielsson (see above, Chapter Eight) and Hofstadter and Dennett (see above, Chapter One) the reader may find Colby, K M, *Artificial Paranoia* (Pergamon Press, Oxford, 1975) of interest.

A summary of personal construct theory is given in Kelly, G A, *The Psychology of Personal Constructs* (Norton, New York, 1955).

Index

INDEX

R*einventing Man* is a journey into the state of the art in robotics as well as an exploration of the possibilities of 'reinventing' human intelligence in machines. Although the modern robot owes its existence to a progression of new technologies—among them, computers, cybernetics, silicon memory chips — more than this is required if we are to create a true thinking robot: human biology, psychology, and even philosophy must come into play. Roboticists now at work on the problem of artificial intelligence are trying to bridge the gap between mind and body. If they succeed, they are likely to shed light on one of mankind's greatest mysteries: the nature of the mind itself.

What are the prospects of creating machine intelligence comparable to our own? While it is a sophisticated piece of machinery, the present-day robot in large part has lacked two very important 'human' qualities: intelligence and perception. Until recently, the robot has been unable to see and react to its environment — to learn — the same way a human being does.

Technological breakthroughs made in the past few years are changing all of this. *Reinventing Man* offers a look into a future that is already here, by introducing WISARD, probably the closest anyone has come to date to creating a thinking, seeing, sensing, self-programming learning machine. WISARD is a sharp break from the traditional idea of programmed computers functioning according to sequential circuits. It is, instead, made up of unprogrammed silicon memory chips, randomly arranged like neural networks in the brain. Here is a machine that, like humans and unlike programmed computers, can actually learn from experience and by induction.